LIFE ITSELF

Complexity in Ecological Systems Series

COMPLEXITY IN ECOLOGICAL SYSTEMS SERIES

T.F.H. Allen and David W. Roberts, Editors

Robert V. O'Neill, Adviser

ROBERT ROSEN

Life Itself: A Comprehensive Inquiry Into the Nature, Origin and Fabrication of Life

T.F.H. ALLEN AND THOMAS W. HOEKSTRA

Toward a Unified Ecology

LIFE ITSELF

A Comprehensive Inquiry Into the Nature, Origin, and Fabrication of Life

ROBERT ROSEN

COLUMBIA UNIVERSITY PRESS • NEW YORK

COLUMBIA UNIVERSITY PRESS
NEW YORK OXFORD

Library of Congress Cataloging-in-Publication Data

Rosen, Robert, 1934–
Life itself : a comprehensive inquiry into the nature, origin, and fabrication of life / Robert Rosen.
p. cm. — (Complexity in ecological systems series)
Includes index.
ISBN 0-231-07564-2
1. Life (Biology) 2. Life—Origin. 3. Biological systems. 4. Biology—Philosophy. I. Title. II. Series.
QH325.R57 1991
574—dc20 91-3110
CIP

Casebound editions of Columbia University Press books are Smyth-sewn and printed on permanent and durable acid-free paper.

Printed in the United States of America

c 10 9 8 7 6 5 4 3 2 1

To the memory of
Nicolas Rashevsky
and
James F. Danielli,
who would have been interested

Contents

Foreword

T.F.H. Allen and David W. Roberts

TWO DEMANDS are being made upon the community of ecologists: that their discipline be increasingly in a predictive mode; and that ecologists be prepared as never before to move up-scale and consider large-scale systems. Leaps in the technology for data acquisition and processing make it feasible to deal with large-scale phenomena in relatively fine grain terms. These imperatives require, and these opportunities allow, ecologists to deal with complex systems. All ecologists are aware that there is much complexity in almost all ecological systems. So impressive is ecological complexity that one becomes convinced that there is something in the very nature of ecological material that is complex. However, it emerges that nature itself is neither complex nor simple. Complexity is a matter of how the observer specifies the system either explicitly or implicitly in the way questions are cast. What makes ecology complex is the challenge of the questions we dare to ask of nature. When the whole entity displaying the phenomenon is scaled much larger than the entities used as explanatory principles, then the system is complex. It is therefore the urgency of certain questions that presses ecologists into the realm of complex systems.

In complex systems, abdication of the scientists' responsibility for specification and scaling leads to confusion and contradiction. Complexity requires a strict and consistent epistemology. The paradigm of this series is that complexity is tractable but demands parallel description at many explicitly specified levels. In the face of complexity it is essential to distinguish model and observables from the material system, and to recognize that the model must invoke a scale and a point of view. With that in mind, the books in this series explore many facets of ecology broadly defined.

The first two books in the series, *Life Itself* by Robert Rosen and

Toward a Unified Ecology by T. F. H. Allen and T. W. Hoekstra, approach theory in the most general terms. The present volume is as much about the general properties of life as it is about ecological systems narrowly defined. Such is the scope of this project that a book about the generalities of living systems is well within the series' purview. By contrast, in preparation are several books that take specific facets of the unified ecology and lay out their complexities. Some volumes in the series will be on distinctly applied topics like economic ecology or the larger issues in environmental regulation. A majority of the volumes will, however, deal with ideas in basic ecology, such as complexity in spatially defined landscape systems and the difficulties of aggregation as one moves between ecological hierarchical levels.

While the theory will employ whatever mathematics is necessary, the editors hope for a style of theory that is understandable because of its intuitive appeal and clear relevance. The series is intended to offer a useful context for specific research projects on particular types of organisms, perhaps in some local setting. The main body and mainstream of ecological research will always have that quality of real organisms or tangible processes in real places. In hierarchical systems, the upper level gives meaning to the level of focus, the lower level. There need be only a small difference between ecological research that is prosaic and isolated as opposed to research in those same areas that deservedly commands attention from pertinent related areas of research, and the ecological research community at large. The editors will try to have the series offer a means of casting mainstream research in a meaningful context. By presenting a milieu in which the mainstream of ecology moves forward, we expect to enable a more coherent ecology, but one performed by specialists within the various subdisciplines as appropriate.

Preface

FOR WHOM is this book intended? I do not know. The book itself has no pragmatic purpose of which I am aware. It is thus for anyone who wants to claim it. My sentiments in this regard are exactly expressed in the words with which Johann Sebastian Bach dedicated the third part of his *Clavierübung:*

Bestehend in verschiedenen Vorspielen . . .
Denen Liebhabern und besonders denen
Kennern von vergleichen Arbeit
zur Gemuths Ergezung.

(Consisting of sundry preludes . . . for those who love such things, and particularly for those who are familiar with comparable work, for the delight of their souls.)

This book represents part of the outcome and present status of about thirty years' work on the problem "What is life?" Some of it has appeared elsewhere, in the form of journal articles, or in some of my other books. But most of it, especially the epistemological considerations at the heart of it, has accumulated unpublished and unpublishable, except in this form.

I must add that writing this text is the hardest thing I have ever tried to do, much harder than doing the research it embodies. The problem was to compress a host of interlocking ideas, drawn from many sources, which coexist happily in my head, into a form coherently expressible in a linear script. Moreover, there was the problem of trying to indicate the richness of many of the ideas, which in themselves want to ramify off in many directions, while keeping them focused on the primary problem. This last has involved a rigorous, and at times exquisitely painful, selection process;

I could easily have expanded almost every chapter, and indeed many of the chapter sections, into separate volumes each the size of the present one.

I have tried to address these problems of organization and exposition in the same way as I approach research, namely, to try to let the problems tell me what to do. Everything to be found herein, and the manner in which it is expressed, is what it is because it seemed to me to be necessary. And about each, I can say in truth that it is the best I can do. It is for the reader to say whether this has been enough.

Note to the Reader

THIS BOOK, like most books, has a primary structure and a tertiary structure. The primary structure is, of course, the linear string of symbols, words, paragraphs, and chapters that is mandated by the nature of the language. But just as with a molecule of globular protein, which must fold in order to become active, this linear string must be folded back on itself, so as to bring parts remote on the string into close contact. This particular volume is very heavily folded indeed. I have tried to indicate at least some of the major folds via an elaborate system of cross-references, sometimes back to parts of the string we have already traversed, and sometimes ahead to parts yet to come. Many other folds are not so indicated, but I hope that enough of them have been specified so that the remainder are obvious.

In any case, an overview of the fully folded structure, and a word about why it is folded as it is, may be helpful at the outset.

First and foremost, it should always be borne in mind that this book is about biology. More specifically, it is a report of where the question "What is life?" has taken me in the quest for an answer. I am well aware that most of the ideas developed herein seem, in isolation, to have little to do with conventional biology. But ideas fold too; even those that seem most remote in terms of their initial origin, and even their content, may turn out to lie very close together indeed in some appropriate topology.

Indeed, at the risk of pushing the folding metaphor too far, this entire volume is an attempt to show how the ideas developed herein constitute, in effect, a single enzyme geared to lyse the problem "What is life?"

It has turned out that, in order to be in a position to say what life *is,* we must spend a great deal of time in understanding what life is *not.* Thus, I will be spending a great deal of time with mechanisms and machines, ultimately to reject them, and replace them with something else. This is in fact the most radical step I shall take, because for the past three centuries, ideas of mechanism and machine have constituted the very essence of the adjective "scientific"; a rejection of them thus seems like a rejection of science itself.

But this turns out to be only a prejudice, and like all prejudices, it has disastrous consequences. In the present case, it makes the question "What is life?" unanswerable; the initial presupposition that we are dealing with mechanism already excludes most of what we need to arrive at an answer. No amount of refinement or subtlety within the world of mechanism can avail; once we are in that world, what we need is already gone. Thus, we must retreat to an earlier epistemological stage, before the assumptions that characterize mechanisms have been made.

The stage to which we retreat is embodied in what we call Natural Law, which seems to me to be the bare minimum required to do science at all. The essence of it lies in what we call the modeling relation. Roughly speaking, this involves only the bringing of two systems of entailment into congruence. Modeling relations provide the thread that ties everything in this volume together.

From this perspective, the hypotheses of mechanism turn out to be only a very special way of embodying Natural Law, and correspondingly, a very limited way. As it turns out, mechanism presupposes the identity of two modes of modeling that can be very different from each other and that we exhibit under a wide variety of rubrics: analytic/synthetic, syntactic/semantic, product/coproduct, etc. Presupposition of their identity thus creates a degeneracy, a nongenericity; it is precisely this very degeneracy that has become identified with science itself. But it is, from my perspective, an impediment; when I remove it, I glimpse a whole new world of possibilities, with exactly the same claim to be called science as mechanism does. Only in such a world, as I argue, do we find the resources to grapple with the question "What is life?"

My arguments will be in no way speculative. At each stage, my conclusions are forced by the nature of the problem itself, and to where I have been led by preceding stages. This procedure in fact leaves no room for speculation at all. In my view, theory is the very antithesis of speculation, despite an all-pervading confusion between the two, an inability to tell a hypothesis from a conclusion.

I shall spend a great deal of time discussing the concept of state and recursive state transitions. These represent the central features of mechanism. Recursiveness is the key property; it connotes a situation in which entailment can in some sense be moved from domain to range. Only very special entailment structures allow this; accordingly, the presuppositions of mechanisms automatically mandate a corresponding impoverishment of entailment. In another language, there is not much causality in a mechanism; almost everything about it is unentailed.

As I shall argue at the end, organisms sit at the other extreme of the entailment specturm than mechanisms do; almost everything about them is entailed by something else about them. This is why an initial presupposition of mechanism is so devastating; it restricts us to fragments, pieces that individually can be regarded as mechanisms all right but that cannot be articulated or combined within those confines.

There are two ways out of the world of mechanism that I shall describe. One of them concerns the circle of ideas called *relational,* which I develop at some length from various points of view. These ideas are concerned with function; because this word "function" has no general mechanistic counterpart, it is accordingly regarded as unscientific or prescientific. Nevertheless, there is a class of mechanisms (called *machines*) that do admit the concept of function and that accordingly have relational descriptions. As I show, already in machines, the relational description involves a divorcing or separation of function (embodied in the relational description) from structure (embodied in the mechanistic ones). Relational descriptions retain meaning even when structural ones do not. Hence relational ideas can provide one kind of exit from the cage of mechanism.

Another exit arises from the application of a familiar formal procedure, the taking of limits. In general, the limit of a sequence of mechanisms need not be a mechanism; the limit of a sequence of mechanical models of a system may still be a model but not a mechanical one. A system of this kind will be called a *complex* system (although this usage of the term "complex" differs somewhat from that of most other authors).

Thus, I come at the end to propose an answer to the question "What is life?" In a nutshell the answer can be expressed this way: an organism is a material system that (1) is complex, and (2) admits a certain kind of relational description.

The consequences of these ideas are indeed radical. In a sense, physics shrinks and biology expands. Physics as we know it today is, almost entirely, the science of mechanism, and mechanisms, as I argue, are very special as material systems. Biology involves a class of systems more general than mechanisms. In fact the relative positions of physics and biology become interchanged; rather than physics being the general and biology the special, it becomes more the other way around. There are other consequences as well, to which I draw attention in the text as I come to them.

This volume concludes, as it must, with a discussion of the origin problem, which I pose in the context of *fabrication.* In complex systems, as

I argue, the chain of entailments involved in fabrication of a system is completely different from that underlying the operation of the system. In mechanisms, on the other hand, owing partly to the causal impoverishment to which I have already alluded, the same kinds of causal chains underlie both. On the other hand, an entirely new feature manifests itself precisely here; because organisms embody so much entailment, a relational theory of organisms is also a general theory of fabrication. These remarks will provide the point of departure for subsequent volumes.

On the basis of Natural Law, then, we are allowed to draw quite a number of drastic conclusions. What Natural Law gives us are certain categories of formal objects, the category of all models of a given material system. Science itself is bound up essentially with the structures of such categories, and the relations that can exist among them (which in turn reflect themselves as relations of metaphor, analogy, etc., among the natural systems represented by the models). Ideas drawn from the Theory of Categories correspondingly permeate my enterprise; it provides the natural language for expressing both my arguments and my conclusions.

This book necessarily involves a lot of formalism, often as subject, often as tool, sometimes as both. But it is not formal in the usual sense; what I am talking about does not admit formal treatment in that sense. My subject matter is rather the role of formalisms in describing entailments in the natural world and what the formalisms *mean* in such a context. Mathematicians, both pure and applied, often find such considerations repugnant, but that is too bad; it cannot be helped.

I conclude these preliminaries with a personal word, regarding the genesis of these ideas reported below. I started from early childhood with a lust to do biology, which I retain. In pursuit of this lust, I acquired a great deal of mathematics; to me, this seemed natural, because as was said before, ideas fold. My main preoccupation in those days was to learn all about operator algebras, the language of quantum mechanics, which I then believed would be sufficient. Almost by accident, I also absorbed a lot of Theory of Categories. It has turned out that the latter was much more important than the former.

My first steps in actually trying to do biology happened to be relational. I was early convinced that such considerations provided legitimate models, descriptions of natural systems that had as much right to be called models as any system of differential equations.

The trouble came when I tried to integrate relational and structural descriptions of the same biological systems. They did not seem to want to go together gracefully. Yet they *must* go together, being alternate descriptions of the same systems, the same material reality. Moreover, I needed both; biology seemed to require both.

They say that all science must start from experience. Mine was that relational models and mechanical models, drawn from physical analysis through reductionism, were not going together. That was a fact. My conclusion from that fact was that I was simply being stupid, or else there were some deep and essential things embodied in that fact. I was never able to rule out the first possibility, but the possibility of the second is what has led circuitously, in the course of time, to what is chronicled herein. In short, this is where I have been led by merely following the problem. It is the problem that imbues the path itself with whatever intrinsic logic is discernable in retrospect.

And so, I repeat, what is in this book is about biology. It directly addresses the basic question of biology: "What is life?" And nothing can be more biological than that.

LIFE ITSELF

I

PRAELUDIUM

SOME WEEKS after completing the manuscript of this monograph, I had to prepare the essay that follows. It constitutes the text of a talk I was asked to give at a scientific conference, and the title was predicated for me in advance. Naturally, like all preludes, it was written in the light of what had preceded it. I believe that, as it turned out, it also constitutes the best possible introduction to the rest of the volume. It is not an abstract or summary in the usual sense; rather, its relation to the text itself is (to use another currently popular image) that of a piece of a fractal to the whole thing.

I believe that by reading this introductory essay, this praeludium, the reader will obtain a clearer idea of the whys and wherefores of what follows than any summary could provide.

"Hard" Science and "Soft" Science

Some years ago, the novelist C. P. Snow drew attention to a dualism that permeates and poisons the intellectual life of our times, a dualism between science and art, between science and humanism.

The dualism to which Snow, among others, drew attention is indeed real. It has always been real and has existed since human beings first learned to think and to communicate their thoughts. But the situation is, and always has been, far worse than Snow has depicted. He painted a picture of science itself as a kind of pure phase, and its relation to other aspects of our culture as a kind of phase separation; scientists and humanists separating from each other as oil separates from water, through a preference of like for like, and an antipathy of like for unlike. But the dualities that Snow depicted also permeate science itself.

I have, much against my will, been immersed my whole life in one of these dualities, namely, the antagonism between "theory" and "experiment." My subject matter herein is another, in fact closely related duality, that between "hard" science and "soft" science, between quantitative and qualitative, between "exact" and "inexact."

This duality is not to be removed by any kind of tactical accommodation, by any superficial effort of conciliation or ecumenicism. The antipathies generated by the duality itself are only symptoms of a far deeper situation, which has roots partly in specific subject matter, partly in individual aspirations, and most important, in the embracing of mutually incompatible *weltanschauungen,* which reflect the deepest aspects of temperament and personality. It is thus not a matter of logical argumentation or persuasion that is involved here; it is a matter more akin to religious conversion.

In what follows, I discuss the duality between qualitative and quantitative. As we will see, in the sciences this dichotomy rests on (generally unrecognized) presuppositions about the nature of material reality and on how we obtain knowledge about it. I will then show that these presuppositions themselves have formal, mathematical counterparts, which allow us to reflect this scientific dualism into an exactly parallel one that exists within mathematics itself. This mathematical form of the dualism is centered around the notion of *formalization;* it can be expressed as the duality between syntactics and semantics; between what is true by virtue of form alone, independent of any external referents, and what is not.

The virtue of doing this is that there is a theorem (Gödel's Theorem) that actually resolves the issue, at least in part. When we pull this theorem back into a scientific context, by looking at its epistemological correlates, we obtain thereby some new and deep insights into the duality between quantitative and qualitative; between "hard" and "soft." I think that all concerned will find some surprises in this exercise.

Naturally, in this brief space, I can give only the most cursory sketch of the ideas involved. But I hope that enough will be said to provoke some reappraisals on both sides of the duality.

QUALITIES AND QUANTITIES IN THE SCIENCES

I can perhaps best illustrate the dichotomy between quality and quantity in the sciences with two quotations. The first is due to Ernest Rutherford:

> Qualitative is *nothing but* poor quantitative.

The emphasis is mine. The second quotation is due to Robert Hutchins, a man no less clever than Rutherford:

A social scientist is a person who counts telephone poles.

I chose these, out of the countless others that could be used, because these words are fighting words; they most vividly exhibit the emotional character of the issues involved.

Obviously, for Rutherford, *everything* we call a quality or a percept is expressible in terms of numerical magnitudes, without loss or distortion. For Rutherford, therefore, every quality can be quantitated, and hence, measured and/or computed. For Rutherford, science does not begin until this quantitation is made, until crude and inexact talk about quality is replaced by precise, exact, and completely equivalent talk about numbers. Indeed, to discuss qualities in any other terms is contemptible ("poor quantitative").

Hutchins tacitly accepts Rutherford's equation of "science" with "quantitative," but for him, this makes the phrase "social scientist" an oxymoron, a contradiction in terms. For Hutchins the features or qualities of a social structure that are of interest or importance are precisely those that are *unquantifiable;* conversely, anything we *can* count is trivial or irrelevant ("telephone poles").

Could anything more clearly exhibit the issues involved here? Let me articulate a few of them. Can arbitrary qualities (the stuff of perception) be equivalently expressed in terms of a certain limited subset of elementary qualities (those we can measure numerically)? Of so, how? If not, what does it mean to have a science of such qualities? What relationships can exist between such sciences (if indeed, any at all)? It is clear that these issues and others like them involve the deepest aspects of the relation between the perceiving mind and the perceptual universe, that the attitudes expressed by Rutherford and Hutchins in the quotations above involve radically different views of these questions, and that they cannot both be right.

Rutherford's view, that every perceptual quality can, and must, be expressible in numerical terms, is associated with the viewpoint commonly called *reductionism.* In practice, reductionism actually asserts much, much more than this; in its most extreme form, it actually identifies a specific family of elementary numerical qualities (and the procedures for measuring them, at least in principle) and anchors them in *physics.* How this is done is a very long story, which I cannot enter upon here but which can be found in an earlier work.[1] According to this view, there *is* no other science than

1. R. Rosen, *Anticipatory Systems* (New York: Pergamon Press, 1985). Hereafter cited in text as *AS.*

physics; everything else we call a science is ultimately a special case of physics.

Hutchins' view, on the other hand, is that physics, in this sense, is itself only a very special science, limited precisely to those very special qualities that happen to also be quantifiable. Accordingly, reductionism is wrong in principle; each science must have its own character and its own procedures, shaped by the specific class of qualities with which it must deal. To Hutchins, who was throughout his life concerned with qualities of social systems that pertain (in the broadcast sense) to politics, it was *precisely* to the extent that something was quantifiable, or expressible in terms of numerical magnitudes, that it was irrelevant. Thus, in effect, Hutchins is inverting Rutherford's dictum; he is asserting that *quantitative is poor qualitative.*

There is nothing inherently illogical, or even unscientific, about either of these positions. They differ so radically because they clearly start from entirely different philosophical presuppositions about the nature of the perceptual world and the relation of the perceiver to the percepts. It is clear that Hutchins and Rutherford could hardly communicate beyond superficialities; they could not be friends. That is precisely the dualism between "hard science," personified here by Rutherford, and "soft science," personified by Hutchins.

There are ways out of this impasse, but they are not palatable to either party. They involve a recognition that mathematics has more to offer besides numbers and a corresponding recognition that perceptual qualities may be expressed in terms of them (i.e., that "measurement" is not the simple thing Rutherford thought it was). But once again, the exploration of these matters is not my present purpose.

Syntax and Semantics

I am now going to do something that would bother both Rutherford and Hutchins, though in different ways. I am going to illuminate the duality they personify by looking at a cognate situation in an entirely different realm, the realm of mathematics. I do this for two separate reasons: first, because it *is* illuminating, and second, because the *fact* that it is illuminating is one that neither Hutchins nor Rutherford could account for.

Let me begin with a few words about the relations existing between the mathematical universe and the perceptual one. It is a fact of experience, for instance, that

$$2 \text{ sticks} + 3 \text{ sticks} = 5 \text{ sticks}.$$

On its face, this is a proposition about *sticks.* But it is not the same kind of proposition as, say, "sticks burn" or "stocks float." It differs from them in that it is also about something else besides sticks, and that "something else" takes us into the world of mathematics.

The mathematical world is *embodied* in percepts but exists independent of them. "Truth" in the mathematical world is likewise manifested in, but independent of, any material embodiment and is thus outside of conventional perceptual categories like space and time. These facts have indeed, from the time of Pythagoras on, spawned another profound dualism, a dualism between idealism (which at root is an attempt to extend the reality of number to the rest of the perceptual universe) and materialism (which is an attempt to include "mathematical reality" inside conventional perceptual realms).

But of this I need not speak. To motivate our discussion, it is enough to observe that both science, the study of phenomena, and mathematics are in their different ways concerned with systems of *entailment,* causal entailment in the phenomenal world, inferential entailment in the mathematical. At root, where Hutchins and Rutherford differ is precisely in their views about entailment, about what is entailed from a datum and about how that datum is itself entailed. Hence, at a sufficiently deep level, the controversy between them, and the dualism they represent, pertains to entailment itself, entailment in the abstract, free of any qualifying adjectives like "causal" or "inferential."

It is in this sense that I turn to the mathematical world in order to illuminate what it tells us about entailment. That is, I will be talking about entailment, rather than about mathematics, just as, in the example above, I could talk about number while apparently talking about sticks.

Mathematics over the past century has given little evidence that it is concerned with eternal, timeless, and hence, unarguable truth. On the contrary, contemporary mathematics is filled with (no pun intended) chaos and turbulence, bespeaking a profound internal instability. Historically, we are still witnessing what (it is hoped) are the transients arising from two profound shocks: the overthrow of Euclid and the discovery of inconsistencies (paradoxes) in Set Theory.

To be sure, most practicing mathematicians, like most practical (empirical) scientists, go on about their business, indifferent to such matters, convinced to the depths of their soul about the reliability of what they do; like corks, they believe they will float on calm and troubled waters alike. But I am speaking of foundations, and it is here we shall look.

The two great shocks of which I spoke above have coalesced, beginning in the early years of the present century, into a frantic concern with *consistency,* with a demand that a system of inferential entailments (e.g., a set of axioms or production rules, operating on a set of given propositions or postulates) be free of internal or logical contradictions. How can we be sure that a system of entailment, e.g., a mathematical system in the broadest sense, is consistent? I shall be concerned with one particular kind of answer given to this question, an answer championed by David Hilbert, which can be summed up in a single word: *formalization.*

Hilbert and others thought they had traced down the ultimate source of all the difficulties in mathematics. They pointed out that propositions in mathematics are nominally *about* something; i.e., they have meanings that involve referents outside themselves. Thus, for instance, in Euclid, the word "triangle" is not *just* an array of letters to be manipulated in a certain way; it *refers* to a rich and vivid kind of geometric object. And even beyond that, it even refers to things in the phenomenal world. In that sense, any Euclidean proposition containing the word "triangle" can be thought of as describing a *percept* or *quality* manifested by this external referent.

Thus, according to this analysis, mathematical truth had come to involve two distinct aspects, one pertaining to how we are allowed to manipulate the word "triangle" from one proposition to another, and another pertaining to the actual referents of that word. I will call the former *syntactic* truth, the latter, *semantic* truth. Hilbert and his colleagues argued that it was precisely by allowing semantic truth into mathematics at all (i.e., in the admissibility of regarding a mathematical proposition as the description of a percept or quality of something, allowing a mathematical proposition to *refer* to something) that all the difficulty arises.

Hilbert and his formalistic school actually asserted much more than this. They argued that what we have called *semantic truth could always be effectively replaced by more syntactic rules.* In other words, any external referent, and any quality thereof, could be pulled inside a purely syntactic system. By a purely syntactic system, they understood: (1) a finite set of *meaningless* symbols, an alphabet; (2) a finite set of rules for combining these symbols into strings or formulas; (3) a finite set of production rules for turning given formulas into new ones. In such a purely syntactic system, *consistency is guaranteed.*

The best statement I have even seen of the formalistic program is that given by Kleene; it does no harm to quote it again:

> (Formalization) will not be finished until all the properties or undefined terms which matter for the deduction of theorems have been expressed by axioms. Then it should be possible to perform the deductions treating the technical terms as words in themselves without meaning. For to say that they have

> meanings necessary to the deduction of the theorems, other than what they derive from the axioms which govern them, amount to saying that not all of their properties which matter for the deductions have been expressed by axioms. When the meanings of the technical terms are thus left out of account, we have arrived at the standpoint of formal axiomatics.[2]

The idea that all truth can be expressed as pure syntactic truth, which is the essence of the formalist position in mathematics, I claim to be the analog of Rutherford's position in science, the formal analog of "hardness" and quantitation.

The formalist position is, first of all, an expression of a belief that all mathematical truth can be reduced to, or expressed in terms of, word processing or symbol manipulation. Hence the close association of formalization with the idea of "machines" (Turing machines) and with the idea of algorithms. These embody purely automatic procedures, which require no thought, no perception, indeed, no external agency at all.

Second, the formalist position, that the universe of discourse needs to consist of nothing more than meaningless symbols pushed around by definite rules of manipulation, is exactly parallel to the mechanical picture of the phenomenal world as consisting of nothing more than configurations of structureless particles, pushed around by impressed forces.

The formalist position seems, on the fact of it, very attractive. For, by asserting that all truth is syntactic truth, it tells us that (1) we lost no shred of mathematical truth in the process of formalization, and (2) we are automatically guaranteed that mathematics is consistent. We pay for these benefits by giving up the idea that mathematics is "about" anything, i.e., that its propositions express percepts or qualities, but on the other hand we are *informally* free to interpret these propositions in any way we want. These are, of course, exactly the same attractions that the "hard" or quantitative sciences offer in the phenomenal world.

GÖDEL'S THEOREM

The celebrated Incompleteness Theorem of Gödel[3] effectively demolished the formalist program. Basically, he showed that, no matter how one tries to formalize a particular part of mathematics (Number Theory, perhaps the inmost heart of mathematics itself), syntactic truth in the formalization does not coincide with (is narrower than) the set of truths about numbers.

There are many ways to look to Gödel's Theorem. Indeed, the Theorem

2. S. C. Kleene, *Metamathematics* (New York: van Nostrand, 1952).
3. K. Gödel, *Monatshefte Math. Physik* (1931), 38:173–198.

itself has provoked an enormous literature, as might be expected. For our purposes, we may regard it as follows: *one cannot forget that Number Theory is about numbers.* The fact that Number Theory is about numbers is essential, because there are percepts or qualities (theorems) pertaining to numbers that cannot be expressed in terms of a given, preassigned set of purely syntactic entailments. Stated contrapositively: no finite set of numerical qualities, taken as *syntactical* basis for Number Theory, exhausts the set of all numerical qualities. There is always a purely semantic residue, that cannot be accommodated by that syntactical scheme.

Gödel's Theorem thus shows that formalizations are part of mathematics, but not all of mathematics. Mathematics, like language itself, cannot be freed of all referents and remain mathematics. Any attempt to do this (i.e., any attempt to capture *every* percept through a formalization of any finite set of percepts) must already fail in the Theory of Numbers.

On the other hand, Number Theory is still mathematics, still a system of inferential entailment in itself. It is only that it is not a purely syntactic system, not entirely a matter of word processing or symbol manipulation, independent of any external referent. In other words, Number Theory is not a closable, finite system of inferential entailment. These facts, as embodied in Gödel's Theorem, do not make us give up Number Theory as a part of mathematics nor even give up formalization as a strategy for studying certain kinds of mathematical systems. They express rather the limitations of formalization; it is not, as Hilbert thought, a *universal* strategy. If mathematics is a war against inconsistency, then that war is simply not as easily won as Hilbert believed.

COMPLEX SYSTEMS

The relation between Number Theory and any formalization of it concretely embodies certain features that bear essentially on the dualism with which we started, between "hard" or quantitative science (which asserts roughly that physics must be everything) and "soft" or qualitative sciences (which assert that physics is nothing).

The first thing to bear in mind is that both Number Theory and any formalization of it are both systems of entailment. It is the *relation* between them, or more specifically, the extent to which these schemes of entailment can be brought into congruence, that is of primary interest. The establishment of such congruences, through the positing of referents in one of them for elements of the other, is the essence of the *modeling relation,* which I have discussed at great length elsewhere.[1]

In a precise sense, Gödel's Theorem asserts that a formalization, in which all entailment is syntactic entailment, is too *impoverished in entailment* to be congruent to Number Theory, no matter how we try to establish such a congruence. There are thus qualities pertaining to numbers, and to Number Theory, that are missed by any such attempt; hence any entailments in Number Theory pertaining to these unencoded qualities are likewise inaccessible in the formalization. It would thus require, at best, an infinite number of distinct formalizations to capture all the qualities, and hence, all the entailments of Number Theory, in terms of syntax alone.

This kind of situation is what I have elsewhere termed *complexity.*[4] In this light Gödel's Theorem says that Number Theory is more *complex* than any of its formalizations, or equivalently, that formalizations, governed by syntactic inference alone, are *simpler* than Number Theory. To reach Number Theory from its formalizations, or more generally, to reach a complex system from simpler ones, requires at least some kind of limiting process.

Rutherford's position, as articulated above, can be rephrased as asserting that *every material system is a simple system.* Indeed, I have shown elsewhere that this position is just another form of Church's Thesis,[5] a direct assertion of the simulability (i.e., the purely syntactic character) of mathematical models of reality (i.e., of systems of *causal* entailments).

To a mathematical Rutherford, then, Number Theory would look *soft* relative to its formalizations, precisely because there are more qualities, and hence more entailments, in Number Theory than could be accommodated in terms of "hard" (i.e., syntactic) entailments. This is exactly why biology looks soft to a physicist, for example. I believe I need not belabor this situation, and the mistaken presumptions it manifests, any further.

On the other hand, let me now observe that the relation between Number Theory and a formalization of it can be iterated. That is: in the discussion so far, I have treated Number Theory as a "system," and formalizations as "models" of it. As we have seen, formalizations, being purely syntactic, have too little entailment to capture all the qualities of Number Theory itself. But as I have said, Number Theory is in itself a system of entailments, not only in itself a perfectly good mathematical system, but in many ways the very center of mathematics. Let us suppose we treat *it* as a model (i.e., as we treated formalizations before) and ask what *it* can model.

The question immediately arises: are there other mathematical formal-

4. R. Rosen, *Int. J. Gen. Systems* (1977), 3:227–232.

5. R. Rosen, *Bull. Math. Biophysics* (1962), 24:375–393; R. Rosen, in *The Universal Turing Machine: A Half-Century Survey,* R. Herken, ed., (Kammerer & Unverzagt, 1988), pp. 523–537.

isms too rich in entailment to be captured by Number Theory, and hence *more complex* than it? The answer, of course, is *yes;* in fact we can iterate this process, obtaining more and more complex formalisms, indefinitely.

At each stage of this iteration, a formalism appearing at that stage would appear "soft" with respect to any formalism appearing at an earlier stage. But of course, at no point do we exempt ourselves from entailment itself; quite the contrary.

On the other hand, a mathematical Hutchins might argue as follows: *because* Number Theory is "soft" relative to its formalizations (i.e., is more complex than they are), it is therefore immune to *mathematics* and must be studied by other means. This is tantamount to abdicating to syntax alone the right to call itself mathematics and declaring thereby that nonsyntactical modes of entailment fall outside the scope of mathematics. The material counterpart of this line of reasoning is to exempt complex systems from the domain of science precisely because they are complex. This view, it seems to me, is equally mistaken.

SUMMARY

As will be clear from the preceding discussion, it seems to me that the duality between "hard" or quantitative science and "soft" or qualitative science rests on an entirely false presumption. It is not in fact a question of Rutherford versus Hutchins, i.e., a question of doing physics or not doing science at all. It is rather a relative question, of simplicity versus complexity.

There is, as yet, no comprehensive investigation of the ideas I have sketched in the course of the discussion above; they are too new. But it seems that such ideas, or ideas like them, are necessary in many ways. I would in particular draw attention to the way such ideas ultimately rest on entailment alone, on systems of entailment in the material world (causal entailment) and in the world of formalisms or mathematics (inferential entailment), and on comparisons or congruences between such entailment systems. I have come to believe that the concept of entailment provides a reliable anchorage for the scientific enterprise itself, and I accordingly recommend it to your attention.

Chapter 1

Prolegomena

THIS BOOK represents a continuation, an elaboration, and perhaps a culmination of the circle of ideas I have expounded in two previous monographs: *Fundamentals of Measurement and the Representation of Natural Systems* (henceforth abbreviated as *FM*) and *Anticipatory Systems* (abbreviated as *AS*). Both of these, and indeed almost all the rest of my published scientific work, have been driven by a need to understand what it is about organisms that confers upon them their magical characterisics, what it is that sets life apart from all other material phenomena in the universe. That is indeed the question of questions: What is life? What is it that enables living things, apparently so moist, fragile, and evanescent, to persist while towering mountains dissolve into dust, and the very continents and oceans dance into oblivion and back? To frame this question requires an almost infinite audacity; to strive to answer it compels an equal humility.

1A. What Is Life?

Ironically, the idea that life requires an explanation is a relatively new one. To the ancients, life simply *was;* it was a given; a first principle, in terms of which other things were to be explained. Life vanished as an explanatory principle with the rise of mechanics, when Newton showed that the mysteries of the stars and planets yielded to a few simple rules in which life played no part, when Laplace could proudly say "Je n'ai pas besoin de cet hypothèse"; when the successive mysteries of nature seemed to yield to understanding based on inanimate nature alone; only then was it clear that life itself was something that had to be explained.

From whence shall such explanation come? To which oracle shall I put the question? The first thought is: to that same mechanics, that same physics, which first exorcised life from the heavens and which has since plumbed the depths of matter, space, and time. Living things are surely material; they are manifestations of matter; surely then the secrets of matter must contain the secrets of life. Surely the physicist, who is concerned with matter in all of its manifestations, will have eagerly striven to translate insights about matter in general into corresponding insight about matter's greatest mystery.

Oddly enough, the physicist, qua physicist, has shown no such eagerness. The historical fact is that the phenomena of biology have played essentially no role in the development of physical thought or in the application of that thought. Why? Mainly, I think, because theoretical physics has long beguiled itself with a quest for what is universal and general. As far as theoretical physics is concerned, biological organisms are very special, indeed, *inordinately* special systems. The physicist perceives that most things in the universe are not organisms, not alive in any conventional sense. Therefore, the physicist reasons, organisms are *negligible;* they are to be ignored in the quest for universality. For surely, biology can add nothing fundamental, nothing new to physics; rather, organisms are to be understood entirely as specializations of the physical universals, once these have been adequately developed, and once the innumerable constraints and boundary conditions that make organisms special have been elucidated. These last, the physicist says, are not my task. So it happens that the wonderful edifice of physical science, so articulate elsewhere, stands today utterly mute on the fundamental question: What is life?

One of the few physicists to recognize that the profound silence of contemporary physics on matters biological was something *peculiar* was Walter Elsasser. To him, this silence was itself a physical fact and one that required a physical explanation. He found one by carrying to the limit the tacit physical supposition that, because organisms seem *numerically* rare in the physical universe, they must therefore be too special to be of interest as material systems. His argument was, roughly, that anything rare disappears completely when one takes averages; since physicists are always taking averages in their quest for what is generally true, organisms sink completely from physical sight. His conclusion was that, in a material sense, organisms are governed by their own laws ("biotonic laws"), which do not contradict physical unversals but are simply not derivable from them.

Ironically, ideas like Elsasser's have not had much currency with either physicist or biologist, although one might have thought they would please both. Indeed, in the case of the former, Elsasser was only carrying one step further the physicists' tacit supposition that "rare" implies "nonuniversal."

The possibility is, however, wide open that this supposition itself is mistaken. On the face of it, there is no reason at all why "rare" should imply anything at all; it needs to be nothing more than an expression of how we are sampling things, connoting nothing at all about the things themselves. Even in a humble and familiar area like arithmetic, we find inbuilt biases. We have, for instance, a predilection for rational numbers, a predilection that gives them a weight out of all proportion to their actual abundance. Yet in every mathematical sense, it is the rational numbers that are rare and very special indeed. Why should it not be so with physics and biology? Why could it not be that the "universals" of physics are only so on a small and special (if inordinately prominent) class of material systems, a class to which organisms are too *general* to belong? What if physics is the particular, and biology the general, instead of the other way around?

If this is so, then nothing in contemporary science will remain the same. For then the muteness of physics arises from its fundamental *inapplicability* to biology and betokens the most profound changes in physics itself. This situation is, of course, nothing new in physics; it happened when physics was mostly mechanics and had no apparent room for accommodating phenomena of electricity and magnetism; it happened again when the combined arsenal of ninteenth-century physics spent itself helplessly in assaulting phenomena of spectra and chemical bonding. But just as today's armies are equipped only to win yesterday's wars, we cannot expect contemporary physics to successfully cope with problems other than those with which it has already coped.

As we proceed, we will find a great deal of evidence, of many kinds, that leads to just such an unfortunate conclusion. And thus, we find another, basic reason why biology is hard; it is hard because we are fundamentally ill equipped. This is a far cry from merely being ignorant; it is rather that we are misinformed.

In any event, the task that physics has shirked devolves next onto biologists, perhaps properly so, since it is central to their every enterprise. What light, then, do biologists shed on the taproot of their own endeavors? In fact, precious little. Indeed, a rather strange and dreary consensus has emerged in biology over the past three or four decades. On the one hand, biologists have convinced themselves that the processes of life do not violate any known physical principles; thus they call themselves "mechanists" rather than "vitalists." Further, biologists believe that life is somehow the *inevitable* necessary consequence of underlying physical (inanimate) processes; this is one of the wellsprings of reductionism. But on the other hand, modern biologists are also, most fervently, evolutionists; they believe wholeheartedly that everything about organisms is shaped by essentially *historical, accidental* factors, which are inherently unpredictable and to which no universal principles can apply. That is, they believe that everything important about life is *not* necessary but contingent. The unperceived ironies and contradictions in these beliefs are encapsulated in the recent boast by a *molecular* biologist: "Molecular biologists do not believe in equations." What is relinquished so glibly here is nothing less than any shred of logical necessity in biology, and with it, any capacity to actually understand. In place of understanding, we are allowed only standing—and watching. Thus, if the physicist stands mute, the biologist actually negates, while pretending not to.

Thus, to ask the question "What is life?" is to find oneself standing essentially alone. But perhaps not entirely; there is yet another oracle to be consulted. That oracle is System Theory, which as yet speaks only in whispers. Insofar as it can be characterized, System Theory is the study of organization per se, divorced from material embodiment, as the form of a statue can be divorced from the marble or as cardinality can be divorced from the things being enumerated. This oracle will at least entertain the question but only when it has been transmuted to a new form: not what is special about life in terms of matter but what is special about it in terms of organization. This is good, because obviously organisms are, in purely material terms, of the greatest diversity. But it is not, by itself, good enough. Moreover, this oracle speaks not of laws or principles, as physics does; as yet it can speak only in parables.

Thus, the question remains, and it is still *the* question: What is life? It commands us to grapple with it and even allows us the luxury of choosing our own weapons for the struggle. But our armory is inadequate; wherever we look, some essential element is missing. Life is material, but the laws framed to describe the properties of matter give no purchase on life. Something is missing here, perhaps something essential for the understanding of matter in general, however much the physicists insist not. Biology has so far spent itself in cataloguing the endlessly interesting epiphenomena of life, but at the heart of it there is still only a gaping void. And the parables of the system theorist cannot as yet be incarnated in material reality. As I said, something is missing, something big, but it is hard to see even the biggest things when they are not there. We can only sense the void of its absence and try to fabricate what is necessary to fill it.

That is what the present book is about. It is about the creation and the application of an armory for a renewed assault on the question of questions: What is life? In the process, we shall find ourselves partly in the world of physics, constructing a language appropriate for a physics of "organized matter," a physics of *complex systems.* We shall also find ourselves partly in the world of the system theorist, developing a language appropriate to "material organization" and thereby clothing that world in a substance and coherence it has largely lacked. We will be in the world of the mathematician, in the world of formalisms and formalizations. And finally, of course, we will find ourselves in the world of biology, to see how far our armamentarium will take us in our struggle with the question: What is life?

1B. Why the Problem Is Hard

As a first step in our assault on the problem What is life? it will be well to get some idea of what we are up against. Specifically, we will try to understand what it is about the problem that has rendered it so refractory to the combined resources of our contemporary scientific wisdom. This will provide one way of sensing the shape of the void we need to fill and at the same time will help set the stage for our further, more technical developments, though we will not be able to reach a true answer to this question until we come to the end.

Let us begin by noting the very form of this question; we are asking *why*. We shall find ourselves asking "why" very often as we proceed. The answer to such a question (and indeed there are in general many ways to answer such a question) is to assert a "because." As we shall see abundantly later, to ask why is to enter the realm of causality, and to propose an answer is to posit something, to make a hypothesis. Although every physicist must believe in causality, this attitude toward positing a "because" was set long ago by Newton, whose proudest assertion was *hypothesis non fingo*. Indeed, as we shall see, causality in contemporary physics has evolved into a very different kind of thing than that originally envisaged by Aristotle, a thing geared essentially to deal with the question "what?" and to provide answers of the form "this."

One of the main reasons the fundamental question "What is life?" is so hard will turn out to be closely associated with such ideas. It will turn out that this question is really a "why" question in disguise, that we are really asking, in physical terms, why a specific material system is an organism, and not something else. Such questions are not congenial to contemporary science. To take a simple example: If we give a physicist, say, a clock, his or her interest will reflexively concentrate entirely on how it works, *never on how it came to be a clock.* We shall see below that the whole formal structure of theoretical science is geared toward the former question and away from the latter. Indeed, it will turn out that a large part of the work we need to do arises from exactly such elementary considerations.

But there are other possible reasons why the problem is hard. One of the most facile, frequently adduced, and fundamentally misleading reasons is to assert that *we simply do not know enough yet* to meaningfully approach the problem; we need more data. That is, difficulty merely presupposes ignorance, that we lack only an appropriate factual basis on which to proceed. It then follows precisely that, because the problem is hard, any attempt to deal with it is *premature*.

It is well to spend a moment discussing this possibility, since it is plausible on at least two counts. First, the fact is that biology itself is extremely young as a science. Indeed, the very word "biology" dates only from about 1800 or so. It is also true that biology has seldom been able to develop, from within itself, the instruments required for probing what actually goes on inside an organism. For example, however important it was to be able to scrutinize the intimate details of organic anatomy, biology had to wait upon the development of a completely independent physical science (optics) in order to fabricate and deploy so basic an instrument as the microscope. Likewise, the field of biochemistry, not to mention its offshoots of molecular biology and molecular genetics, could not have developed beyond the crudest of generalities without techniques for tracing, manipulating, and characterizing minute quantities of matter—techniques that became available only through developments in atomic and nuclear physics in the 1930s.

Thus, biology has generally had to parasitize other sciences in order to develop its own experimental techniques, and consequently, has barely begun to accumulate its fundamental data. Extrapolating from this historical picture, we may expect that, as new experimental techniques are generated elsewhere, entirely new biological probes, and correspondingly new data, presently unimaginable, await us in the future.

Therefore, just as it was premature to speculate on organic microanatomy before microscopes were available, or to speculate in 1920 on how proteins were synthesized, so too it may be premature to venture upon the much deeper question of "What is life?" today. And the argument can be strengthened by looking at the history of physics itself. Of what value were speculations regarding the source of atomic spectra, or chemical bonding, or the origin of stars in 1850? These were hard problems then, precisely because the factual basis to deal with them was lacking.

Such arguments cannot be discounted a priori, of course, but they can be considerably weakened by precisely the same historical considerations that give them their apparent weight. For one thing, the two most important ideas in contemporary biology, namely, Mendelian genetics and Darwinian evolution, were both premature, according to these views. For another thing, a fact or datum, by itself, is essentially meaningless; it is only the *interpretation* assigned to it that has significance. Thus, for example, one can literally see the rotation of the earth on any starry night; it has always been patently visible, but for millennia human beings did not know how to understand or interpret what they were seeing. Examples of such misinterpretation, which have retarded the development of science by centuries, can be multiplied without end in the history of science; in all these cases, it was the absence, not of data, but of imagination that created difficulty.

Finally, and most important, a fact or datum cannot, by itself, answer a question "why?"

To sum up: It may perhaps be true that the question "What is life?" is hard because we do not yet know enough. But it is at least equally possible that we simply do not properly understand what we already know.

Let us turn to another circle of historical ideas that bear on why biology is difficult. If we look back at the history of physics, with a view to understanding why it could develop as early and as rapidly as it did, we find two relevant sets of circumstances. First, there was the patiently accumulated data of astronomy, collected and tabulated for millennia. Second, an apparently unrelated circumstance: one could experiment directly with simply mechanical contrivances, with inclined planes, pulleys, and springs, and express the resulting data in the form of simple mathematical rules or laws. It was tacitly assumed that these *same* rules, revealed by simply laboratory experiments with simple bulk systems, held good throughout all of nature, at every level, from the very greatest to the very smallest. The story of Newton and the falling apple is perhaps apocryphal, but there is no doubt that his ideas about gravitation were based on an assumption of the universal validity of mechanical extrapolations whether the objects involved were planets, or terrestrial projectiles, or ultimate atomic particles.

This tacit belief in the unlimited uniformity of mechanical behavior, and the corresponding universality of mechanical laws, provided the absolutely essential nutrient that permitted theoretical physics to develop as it did. For it implied that we could study, on a convenient terrestrial scale, the same forces that moved the planets on the one hand and the atoms of the other. It implied that we could, through the simplest laboratory situations, simultaneously study *all* of inanimate nature, that our humble laboratories were proxies for the entire universe.

We now know, after three centuries, that this assumption of uniformity was entirely, hopelessly false. We cannot extrapolate from bulk matter to what goes on in an atom, or from either of these to events on galactic scales. But luckily, it was correct enough, for long enough, to establish a firm foundation of physical thought, a heritage of mathematical language, that could then be infinitely modified in detail and in interpretation, so that when the falsity of the basic premise was exposed, we did not need to jettison everything and start again. In retrospect, we can see that physics has, in this respect, been blessed with unbelievable good fortune.

In biology, however, the situation has been entirely different. From the physical point of view, even the simplest system one would want to call an organism is already inconceivably complicated. There are no biological counterparts of the inclined plane or pulley, the simple system that already manifests in itself the general laws we want to study. We cannot thus study organisms by inorganic proxy, at least not experimentally (see *AS*). The best we can do is to ***dismember*** an organism, to break it apart, and treat its parts in isolation as proxies for parts ***in situ.***

These facts have decreed a very different historical development for biology vis-à-vis physics. They have also served to isolate biology for centuries from the rest of natural science, an isolation to which I have already alluded. It is in fact only in recent years, the years of "molecular biology," that any direct contact at all has become visible between the triumphant ideas of seventeenth-century particle mechanics and the properties of organisms. It is most ironic that today's perceived conjunction between physics and biology, so fervidly embraced by biology in the name of unification, so deeply entrenched in a philosophy of naive reductionism, should have come long past the time when the physical hypotheses on which it rests have been abandoned by the physicist.

The circumstances sketched above provide another complex of reasons why biology is hard and why the ultimate question of "What is life?" is perhaps the hardest of all. But there is one more circumstance that is not without an irony of its own. So far in this section, I have dwelt upon fundamental differences between organic systems and other kinds of material systems, differences that have, as we have seen, essentially set biology apart from the growth of other sciences during the past centuries, and hence, from the canons of explanation and understanding that seemed to work so well everywhere else. From these considerations, one would expect that such vast differences could themselves be quantified, that there should at least be some explicit, tangible, categorical test for distinguishing between a material system that is an organism and one that is not.

Indeed, many people have tried very hard to produce such criteria for separating the quick from the dead. Put briefly, they have all failed. This fact is most significant in itself, but perhaps of even greater significance is the further fact that we can somehow ***know,*** with certainty, that they have indeed failed.

These facts (let us assume that they are such for a moment) mean that we do not even know what biology is *about,* in the same sense that we know what mechanics is about, or what optics is about, or what thermodynamics is about. We thus do not know the scope or the domain of biology, for it has as yet no objectively definable bounds. In place of these, we have only a tacit *consensus,* which we somehow all seem to share, much as the native speakers of a language share the ability to discern its idioms. It is precisely against this tacit consensus that we judge the efficacy of a purported definition of organism, and it is by virtue of this consensus that we know all such definitions have in fact failed. They have failed because all of them, in one way or another, present to us as organisms objects that our consensus rejects, and equally bad, it classifies as nonorganisms systems that our consensus accepts.

The crux of the matter is that, when one tries to embody our recognition criteria for organisms in an explicit list, we find nothing on that list that cannot be *mimicked* by, or embodied in, some patently inorganic system. That is, there is no property of an organism that cannot at the same time be manifested by inanimate systems. Conversely, a dead organism is as inanimate as anything.

Indeed, a whole approach to biology, which used to be called *biomimesis,* was based on this observation. This approach might better be called "artificial biology." The idea was that a material system that could be made to embody "enough" properties of organisms would then *automatically be* an organism. Thus, people studied the motility of oil drops in salt solutions, the irritability and excitability of artificial membranes, the spontaneous division of growing droplets, and a host of other such phenomena. These ideas persist today in at least two areas: experimental approaches to "origin-of-life" and, in an entirely different material context, in robotics. Exactly the same idea is embodied in "Turing's Test"; a "machine" that has enough specific attributes of intelligence must *be* intelligent. From this an entire industry has arisen, devoted to the study of "artificial intelligence."

However we arrive at our consensus concerning what constitutes an organism, we do not implement it by checking off against a list. It may help to regard it rather as a form of "pattern recognition," but this only raises the question of what the pattern is that is being recognized. In any event, the fact remains that biology is not defined by, or characterized by, any discrete list of attributes that collectively reflect the consensus lying at the root of it. There is thus, in particular, no way we can impart that consensus to a "machine" in the form of a program. And as we shall see (see chapters 8–10), this fact itself, when appropriately articulated and judiciously combined with some of the other matters raised above, will take us a surprisingly long way.

I have already mentioned the word "machine" twice in the course of the past few paragraphs. It is a word that comes up often in contemporary biology, especially since almost all biologists account themselves as *mechanists*. Indeed, the protean concept of the machine, which in its way is as ill defined as that of organism itself, has dominated the inner world of biologists for far too long. The specter of the machine constitutes yet another reason why biology is hard. Let us then proceed to discuss it a bit, at least in a preliminary way; I shall have much more to say about it as we proceed.

1C. The Machine Metaphor in Biology

As we have just seen, one of the reasons biology is hard is that no one can say what an organism is. It is, however, all too easy to say what an organism is *like*. In itself, this is not a bad thing to do; trouble arises when one substitutes the latter for the former.

The earliest and most mischievous instance of this kind of substitution goes back to René Descartes. Apparently, Descartes in his youth had encountered some realistic hydraulic automata, and these had made a great impression on him; he never forgot them. Much later, under the exigencies of the philosophical system he was developing, he proceeded to turn the relation between these automata, and the organisms they were simulating, upside down. What he had observed was simply that automata, under appropriate conditions, can sometimes appear lifelike. What he concluded was, rather, that *life itself was automaton-like*. Thus was born the machine metaphor, perhaps the major conceptual force in biology, even today.

Descartes took this fateful step with only the haziest notion of what a mechanism or automaton was (Newton was still a generation away), and an even dimmer notion of what an organism was. But Descartes was nothing if not audacious. Descartes' conception was in fact perfectly timed; the triumphant footsteps of Newtonian mechanism were right behind it; the apparently unlimited capabilities of machines were already on their way toward a complete transformation of human society and human life. Why indeed should the organism not be a machine? There is no denying the many powerful allures encapsulated in the Cartesian metaphor; it hath indeed a pleasing shape.

Aside from its purely scientific and methodological implications, the psychological appeal alone of the machine metaphor to biologists over the years has been immense. We have already noted the profound isolation of biology from the dramatic developments in physical science since the time of Newton. The idea of the organism as machine permitted at least a vicarious contact with all this; it was plausible, easy to grasp, and above all, *scientific;* it showed a way around hazy metaphysical notions like intentionality, *telos,* finalism, and the like, which have always plagued biology, but which physics had apparently discarded for centuries past.

Today, I dare say that the molecular biologist, above all, regards this field as representing the ultimate incarnation of the Cartesian automaton. The molecular machine has now displaced all earler metaphorical images of the organism as clockwork, as engine, as chemical factory, as servo-mechanism, and as computer. Genetic engineers, who are the molecular biologists turned technologues, habitually regards their favorite organism, *E. coli,* as a simple vending machine; insert the right token, press the right button, and the desired product is automatically delivered, neatly packaged and ready for harvest.

But above all, the machine metaphor (supported, of course, by the corpus of modern physics) is what ultimately drives, and justifies, the reductionism so characteristic of modern biology. For whatever else a machine may be, it is a composite entity; it is made up of parts. The way these parts interact to produce the machine's characteristic behaviors constitutes its *physiology;* the way the machine is assembled from these (very same!) parts accounts for its *origin.* The way to learn about these things is merely a matter of dissection or fractionation, to dismember the machine into its constituent parts and characterize them individually as independent subsystems. That is what we do with a machine, be it a watch, or a piece of electronic equipment, or an automobile. It is also what a physicist does with a stone, a molecule, or an atom. All of these procedures embody the first rule of scientific analysis: separate mixtures into pure substances, devolve the properties of the whole onto the properties of the parts.

For the machines we build for ourselves, there is indeed a "set of parts" into which the machine can be resolved; these are obtained essentially by reversing the process by which the machine was fabricated in the first place. These parts can be separated and characterized, without losing any information pertinent either to the physiology or the reconstruction of the machine itself. The belief in reductionism, buttressed precisely by the machine metaphor, extrapolates these facts back to the entire universe; there is *always* a set of parts, into which *any* material system (and in particular, any organism) can be resolved, *without loss of information.* Specifically, the properties of these parts, considered in isolation, collectively entail the properties of the fully articulated system; moreover, they even imply how the parts themselves are to be articulated. In short, the posited set of parts constitutes the natural analytic subunits for simultaneously resolving two *entirely different problems:* how the system is to be constructed (the fabrication problem) and how the system actually works (the physiological problem).

Of course, not every way of decomposing or fractionating a machine will give us a set of parts with these happy properties. Taking a hammer to a watch, for example, will give us a spectrum of parts all right; these may be separated and characterized to our heart's content, but only by a miracle will they tell us either how a watch works or how to make one. This is because two things have happened: application of the hammer has *lost* information about the original articulated watch, and at the same time, it has *added* irrelevant information about the hammer. What the hammer has given us, then, is not so much a set of parts as a set of artifacts.

Nevertheless, the machine metaphor is what whispers to biologists that a set of parts exists; they whisper to themselves that they have found them, with respect to physiology if not fabrication. And yet, we must realize that physicists have long known about very simple material systems, whose "physiologies" cannot be analyzed in this way, even in principle. One cannot, for instance, solve a three-body problem by breaking a three-body system (e.g., the earth, sun, and moon) into three one-body systems or even into a two-body and one-body system. Already in this case, we must draw one of two conclusions: (1) there *is* no set of parts, or analytical subunits, from whose properties in isolation the physiology of the intact system can be reconstructed, or (2) if there are such parts, they are not the obvious ones; they must rather be of a completely different and novel character. Moreover, in this last case, we cannot expect such parts to solve both the physiological and the fabrication problem simultaneously. Instead, entirely separate analyses are required, as befits the two entirely different kinds of problems to be solved.

To sum up: the role of the machine metaphor in biology today is as follows. First, it assures biologists that their subject is an analytical one, because it asserts that any machine is a set of parts. Second, it assures them that the *same* set of parts will solve all problems of fabrication and of physiology simultaneously. Third, it assures them that nothing happens in biology that is outside the ken of the physical universals (or rather of those fragments of physical universality necessary for the understanding of machines). As to the parts themselves, biologists used to think that they were cells, but today they are molecules. And if biology is hard, it is simply because there are so many parts to be separated and characterized.

This last paragraph encapsulates, I think, the working biologist's view of reductionism. If the machine metaphor, which is its primary mainstay, is even a little bit wrong, then this metaphor itself makes biology infinitely harder than it needs to be. It makes biology objectively harder, because it transmutes biology into a struggle to reconcile organic phenomena with sets of constituent fragments of unknown relevance to them; it makes biology subjectively harder because biologists have committed themselves to the analysis rather than to the organism. The question "What is life?" is not often asked in biology, precisely because the machine metaphor already answers it: "Life is a machine." Indeed, to suggest otherwise is regarded as unscientific and viewed with the greatest hostility as an attempt to take biology back to metaphysics.

This is the legacy of the machine metaphor. I hope to convince the reader, in the course of the present work, that the machine metaphor is not just a little bit wrong; it is entirely wrong and must be discarded.

Chapter 2

Strategic Considerations: The Special and the General

I HAVE ASSERTED, several times in the course of the previous discussion, that "something is missing" from the resources we bring to bear on the question "What is life?" The present section is concerned with making this assertion sharper. In the course of doing this, I shall describe a few of the numberless examples of how, in the past, science and mathematics have stumbled into, or more often, been pushed by the pressure of circumstance, into reluctant realization that "something was missing" from their inherited worldview. Such occasions have almost invariably been associated with pain, and the remedies dramatic. Indeed, as true in science as in any other field of human endeavor are the words that George Bernard Shaw puts in the mouth of Andrew Undershaft: "You have learned something. That always feels, at first, as if you had lost something."

2A. Basic Concepts

I shall begin with a brief consideration of the concepts of generality and universality. Specifically, what does it mean to say that a theory is general or universal? Or that a proposition is generally or universally true? It is true that, say, Set Theory is more general than topology, or that Lattice Theory is more general then Group Theory, or that mechanics is more general (universal) than thermodynamics? Or that physics is more general (or less general) than biology? What sense, if any, can we give to such questions?

Let us look at the first question: is set theory more general than topology? At first sight, this question admits contradictory answers. For a topological space is a set, plus additional structure. We might say that the additional structure we need to specify makes topology more special. On the other hand, every set *is* a topological space, if we imagine it equipped with the trivial (i.e., very special) topology. From this point of view, we could equally say that the topological space is the more general concept (since it generalizes the *topological* structure that inheres in every set).

In fact, the question I have raised does not admit resolution at this level; the decision point lies somewhere else. It lies in the fact that, once we have imposed an additional structure on a set, be it of an algebraic, topological, or any other character), the *mappings* we use to compare one such structure with another must, of course, respect that structure. In our case, topological structures are compared by using mappings that are *continuous.* The set of *continuous* mappings from one topological space to another is thus generally much smaller than the totality of unrestricted mappings between their underlying sets. It is at this level that the restrictive nature of the additional structure manifests itself. And on this basis, we have grounds for asserting that set theory is the more general.

On similar grounds, we can argue that mechanics is more general than thermodynamics. Or rather, we can argue that the *formalism* of mechanics is more general than is that of thermodynamics. On this basis, we can assert then that mechanics is the more *universal* theory; I shall have much more to say about this subsequently (see chapter 4).

Comparing *different* kinds of mathematical structures, as for example, a group structure with a lattice structure, requires still another set of ideas, if we wish to talk about the generality of, say, group theory versus lattice theory. What we need to do in this case is to produce one kind of structure in the context of the other, and conversely. For instance, the set of all subgroups of a given group forms a lattice; the set of all automorphisms of a lattice forms a group. If we could do this in such a way that the resulting pair of operations were inverses of one another (e.g., if we could assert, which we cannot, that "the group of lattice automorphisms of the lattice of all subgroups of a group is the group itself" and the corresponding inverse proposition), then we would have grounds for saying that the two theories are equally general; if we could say one and not another, we could rank one theory as more general; if we can say nothing at all, then the generality of the two formalisms cannot be compared; they are of *different* generality.

These rather vague ideas regarding generality serve as a kind of taxonomic index, according to which formalisms may be classified. As we shall see later (see section 3H below), when I come to talk about modeling of one kind of system in another, they actually do much more than this. But for the moment, it is sufficient to note that generality rests on the idea of *inclusion;* a formalism (e.g., set theory) is more general than another (e.g., topology) if there is some kind of proper inclusion that can be established somewhere (in this case, the set of continuous maps between toplogical spaces is a proper subset of the set-theoretic maps between their underlying sets).

In any case, passage from general to special corresponds to the imposition of additional structure, additional conditions not satisfied by the typical (or to use a more modern word, by the *generic*) member of the more general class. Conversely, passage from special to general involves the removal or waiving of such special conditions.

Let us now consider a few well-known historical examples.

2B. From General to Special

From the foregoing, we can see that given any formalism, say the Theory of Groups, we can *generalize* it by simply throwing away some of its defining structure. For instance, if we throw away the group axiom requiring every element to have an inverse, we get a more general structure called a *semigroup* or *monoid.* Here the inclusion on which the assertion of generality rests is patent; every group is a fortiori a semigroup, but not conversely. Indeed, it is not hard to see that, in some precise sense, it is most unusual *(nongeneric)* for a semigroup to *also be a group* or even for an *element* of a given arbitrary semigroup to possess an inverse.

Conversely, the way to pass from a more general situation, (e.g., semigroups) to a more special one (e.g., groups) is to *add conditions.* Namely, we must impose further structure, which serves to discriminate between what is special and what is not. Obviously, the more restrictions we impose, the smaller will be the class of systems that satisfy them all. On the other hand, because there are more conditions to be satisfied, we can usually say more things, and deeper things, about those mathematical objects that do satisfy them. In mathematics, we must strike a balance between these two conflicting aspects of specialization; namely, we must impose enough conditions to allow interesting theorems about the objects satisfying them, but not so many conditions that the class of such objects is too small to itself be interesting. This kind of balance is in fact the mathematical paradise, where one can prove interesting things about interesting objects. We have, of course, seen this before (see section 1A above); it is why physics regards biology as "too special" to be of physical interest.

We have seen that, *given* a formalism, we can specialize it by simply adding appropriate further conditions. But what if we do not exactly have a formalism to begin with? This kind of situation has occurred many times in mathematics, and in its most serious form has been responsible for most of the so-called Foundation Crises that mathematics has experienced over the years. In these cases, some kind of apparently allowable operation or procedure was "too general," in the sense that its unrestricted use implied terrible things like "1 = 0." The question became one of *how* to restrict or circumscribe the procedure, to specify conditions under which it became safe.

A classic example of this situation is the following. The ordinary arithmetic operations (addition and multiplication) are essentially *binary* operations; they apply to pairs of numbers and yield a definite sum and product respectively. They can immediately be extended from pairs of numbers to any *finite* set of numbers; we can give a unique and unambiguous meaning to expressions like

$$r_1 + r_2 + \ldots + r_n; \qquad r_1 \times r_2 \times \ldots \times r_n.$$

Every child knows this.

A whole new mathematical universe was glimpsed through removal of the restriction of finiteness, i.e., by *generalizing* from finite arithmetic to infinite sums and infinite products. In the hands of people like Euler, the Bernoullis, and others, a host of new and beautiful relations were generated; for instance

$$1+\frac{1}{9}+\frac{1}{25}+\ldots+\frac{1}{(2n-1)^2}+\ldots=\frac{\pi^2}{8};$$

$$\sqrt{\frac{1}{2}}\cdot\sqrt{\frac{1}{2}+\frac{1}{2}\sqrt{\frac{1}{2}}}\cdot\sqrt{\frac{1}{2}+\frac{1}{2}\sqrt{\frac{1}{2}+\frac{1}{2}\sqrt{\frac{1}{2}}}}\ldots=\frac{2}{\pi};$$

$$\frac{2}{1}\cdot\left(\frac{4}{3}\right)^{1/2}\cdot\left(\frac{6}{5}\cdot\frac{8}{7}\right)^{1/4}\cdot\left(\frac{10}{9}\cdot\frac{12}{11}\cdot\frac{14}{13}\cdot\frac{16}{15}\right)^{1/8}\ldots=e.$$

It seemed as if the ordinary laws of finite arithmetic extended benevolently into this new, infinite realm: products distributed over sums; the result of an infinite summation or multiplication was independent of how it was parenthesized, etc.

The mathematical masters never (at least publicly) encountered any difficulties in manipulating infinite sums and products. But in the lesser hands that strove to emulate them, patently absurd results began to appear. The worst of them is this:

$$1 + (-1) + 1 + (-1) + \ldots$$

It allows us apparently to conclude that any integer is equal to any other.

Obviously, the world of unrestricted infinite sums and products is somehow *too* large, *too* general. *Sometimes* we can safely navigate in this world, but when? We could, of course, stay completely safe and never enter this world, but that would be too drastic and unacceptable; who would want to renounce beautiful results of the kind we exhibited above? The question became then how to *restrict* ourselves to those infinite sums and products that "make sense" and avoid those that do not.

The resolution of this crisis was given by Cauchy, who in 1805 introduced the necessary concept, *convergence.* He even gave an effective criterion for convergence, which anyone can apply. Thus, the original generalization from finite arithmetic to infinite had to be re-restricted in order to be meaningful; we need, not the universe of *all* infinite sums, but rather all those that satisfy the additional property that they *converge.* And we may note that, as usual, it is *nongeneric* for an infinite sum or product to also converge.

Historically, Cauchy's conception of convergence, which gives sense to limiting operations, and the associated concept of *continuity* (mappings that, in some precise sense, commute with these limiting operations) provided one of the basic cornerstones for what is today called *topology.* A topological space can be regarded as a minimal formalism in which convergence and continuity can be manifested. We may note, for future reference, that present-day topology has other historical antecedents, coming from geometry and dominated by quite different ideas related to congruence. In a way, it is the clash between these two distinct inheritances of topology that has given rise to the Theory of Categories; see section 5I below.

We are, at the moment, still in a Foundation Crisis, arising from the paradoxes inherent in Cantorian Set Theory. These paradoxes (or at least, the ones we know about; there is no guarantee that others may not be lurking), which were discovered around the turn of the century, clearly arise from an unrestricted (too general) use of concepts like membership and elementhood. Particularly suspect is the *reflexive* use of these concepts. As with the earlier problems associated with infinite sums and products, *sometimes* these concepts are meaningful and proper; sometimes they are not. Suffice it to say here that, as yet, no set-theoretic Cauchy has emerged to settle the situation, to find the level (or at least *a* level) of generality that is just right for Set Theory.

2C. From the Special to the General

In the preceding section, we saw some instances where mathematics has proceeded from the general to the special. Most of them were hygienic or therapeutic; in the absence of other exigencies, mathematicians always strive to go the other way, from what is special to what is general. Mathematics seeks the smallest set of conditions under which mathematical truth obtains, because this is the most elegant, the most parsimonious, the most illuminating, in short, the most beautiful.

Sometimes, however, the normally congenial act of generalization in mathematics is associated with great pain. The classic example is the discovery of the "non-Euclidean geometries," dating from about 1820. For over a millennium before that, "geometry" meant one thing; it meant what was between the covers of Euclid's *Elements*. The impact of Euclid upon mathematics in those days is hard to appreciate today; Euclid was the touchstone, the one part of mathematics that had attained absolute perfection. It embodied, in the truest way, the Platonic ideal of mathematics itself. And to use the phrase "non-Euclidean geometry" was at that time simply the most blatant contradiction in terms.

The crisis surrounding geometry that mushroomed in the ninteenth century grew from a single small blemish in the *Elements,* which the Greeks themselves, with their keen eye for such things, had already perceived. The blemish concerned the notorious Postulate of Parallels. The Greek post-Euclidean geometers did not doubt its "truth"; they merely felt it should be a theorem, because, with its potential for the unrestricted prolongation of lines, it was not as "self-evident" as a postulate should be. With their keen nose for aesthetics, they believed that the Parallel Postulate was in fact, redundant, that it was derivable from the other Euclidean axioms and postulates.

As it turned out, of course, the Parallel Postulate was not redundant; it was independent of the rest of Euclidean geometry. Thus, one could get different "geometries" simply by replacing Euclid's postulate with another. Hence, Euclidean geometry lost its unique character; it became only one geometry among many; it became *special.* In fact, it turned out that there was a single number (curvature) that characterized these geometries; Euclidean geometry was flat (curvature = 0); the Lobatchevskian or hyperbolic geometries were those with negative curvature; the Riemannian or elliptic geometries had positive curvature.

This kind of generalization (from zero curvature to nonzero) was of course highly traumatic under the circumstances and in fact was one of the wellsprings of the urgent concern with axiomatics (especially, of consistency of axiom systems), which fed into the Foundation Crises we have already mentioned. It was no longer, for example, even clear what "geometry" was any more. Within less than a half century, geometry had gone from being the most secure part of mathematics to the most insecure. It was a situation in which mathematicians were, for once, not seeking to generalize but were forced to do so.

At roughly the same time, very similar developments were occurring in theoretical physics. In mechanics, it had long been known that the equations of motion of a Newtonian system of particles (at least, a conservative system) were invariant under the Galilean group. This meant that different observers, moving relative to the system they were observing, and to each other, would see the same equations of motion, even though their measurements of position and momentum of particles in the system, tainted by their own motions relative to the system, would be different. In 1905, Einstein pointed out that these same Galilean transformations do not leave Maxwell's equations invariant. Thus, different observers of electrodynamic phenomena would not only come up with different data; they would also come up with different laws of motion. This, to Einstein, was contrary to experience. He showed that Maxwell's equations were invariant to a different group, the Lorentz group. The elements of this group contain a *parameter* c, and in fact, the Lorentz group collapses to the Galilean in the limit $c = \infty$ (or equivalently, when velocities v are small enough so that the ratio v/c is essentially zero). This parameter c was, of course, identified with a fixed (large) number, the velocity of light in vacuum. On plugging the Lorentz group back into mechanics, Einstein was able to draw a number of breathtaking conclusions, which are by now familiar to all.

This *Special Relativity* of Einstein constitutes a generalization of Newtonian mechanics, which came from a completely unexpected quarter. In replacing the Galilean group by the Lorentz group, we are actually doing much the same thing as changing geometries, replacing a flat or Euclidean world ($1/c = 0$) with a curved one; this kind of idea was later extended, by Einstein himself, in the passage from Special to General Relativity.

At roughly the same time (in 1900) another kind of generalization of Newtonian mechanics was in the works, stemming from Planck's discovery of the quantum of action. The nature of this generalization was not, however, fully appreciated for another quarter century, until the development of wave mechanics by Schrödinger, and the mathematically equivalent but very different-looking quantum mechanics of Heisenberg. In particular, the Uncertainty Relations on which quantum mechanics was based contained a (small) parameter h (Planck's constant), and quantum mechanics collapsed back to the Newtonian if we put $h = 0$ (i.e., all observables *commute*).

In fact, this generalization of Newtonian by quantum mechanics is closely related to one that had appeared more than three centuries earlier, namely, the generalization of geometric optics by the wave optics of Huyghens. By the middle of the nineteenth-century, W. R. Hamilton had established his Mechano-Optical Analogy (see *AS*), which allowed him to recast classical Newtonian mechanics in terms of a minimum principle (Least Action), and in the process established an exact dictionary between mechanics and geometric optics. By pursuing this analogy one step further and asking for the generalization of *mechanics,* which corresponded to the Huyghens' generalization of geometric optics, he would have discovered Schrödinger's equation; that is the way Schrödinger himself did it, a very long time later.

Let us look at one further instance of generalization, which we will consider in a mathematical context, but which historically has ties to engineering, and to quantum mechanics as well. These are associated initially with the name of Heaviside. At a time when mathematicians considered discontinuities to be pathological, and when they also thought they knew all about linear systems, Heaviside proposed using discontinuous signals (e.g., step functions) as an entirely novel probe of their behavior. Even worse was his introduction of derivatives of these discontinuities, to obtain things that mathematicians did not consider functions at all. Heaviside's revolutionary approach was thus completely discounted as ugly and as flying in the face of all accepted standards of rigor. His ideas received some prominence when they were adopted by the physicist Dirac, in the course of his development of an elegant formalism for the quantum theory itself (which is as linear as one could want, and where discontinuities are of the essence). However, even so great and so knowledgeable a mathematician as von Neumann completely dismissed Dirac's approach; his remarks are worth quoting directly:

> The method of Dirac . . . in no way satisfies the requirements of mathematical rigor, even if these are reduced in a natural and proper fashion to the extent common elsewhere in theoretical physics . . . the method . . . requires the introduction of "improper" functions with self-contradictory properties . . . the correct structure need not consist in a mathematical refinement and explanation of the Dirac method, but . . . requires a procedure differing from the very beginning. . . .

Nevertheless, within a little over a decade, these supposedly self-contradictory objects were shown to be perfectly rigorous and respectable. They were identified not with functions but with linear functionals, essentially with definite integral operators on linear spaces of functions. Ordinary ("proper") functions turn out to be a very special case, and the more general objects (originally called *distributions*) are identifiable with limits of sequences of these "proper" functions, just as real numbers are limits of sequences of rationals. Thus, from this point of view, distributions *generalize* ordinary functions; the generalization embeds ordinary functions in a new and larger universe, in which they are nongeneric indeed.

The repeated association of limiting processes with generalization of something, of which we have seen a number of instances so far, is itself a very general thing. Limiting processes are a gateway that can take us from a given world to a generally much larger world (since there are more sequences than elements), as we have already seen. The elements of this larger world may have new and different properties from those with which we started; properties generic in the large world but vacuous in the small one that gave rise to them. These remarks should be kept in mind, especially in light of what I have said about the relation of physics and biology. I will not use them for a long time (see chapter 10), but they should not be forgotten.

2D. Induction and Deduction: A Preliminary Note

The foregoing remarks regarding generalizations and specializations illustrate the kinds of ideas we shall need later, when we come to compare biology and physics, organisms and machines. They are, however, a far cry from what most people might expect a discussion of generals and particulars to be about, namely, about induction and deduction. Induction, roughly, seeks to establish general (i.e., quantified) propositions on the basis of *instances;* deduction, conversely seeks to establish instances in terms of quantified or general propositions. Such ideas turn out to have some relation, albeit remote, to the matters discussed above; however, since we shall need some ideas pertaining to induction later, it seems appropriate to say a few words about these issues here. Induction in particular is a very old problem, and the following remarks are intended to be in no way exhaustive.

To fix ideas, let X be some set, whose elements will be called *instances.* Let us further suppose, for simplicity, that we can enumerate the elements of X in some effective way; i.e., we can write

$$X = \{x_1, x_2, \ldots, x_n, \ldots\}.$$

We shall further denote by P some predicate or property that may or may not be manifested by a particular instance x_i in $X;$ if the property P is possessed by the instance x_i, we shall write $P(x_i)$.

Let us denote by x an indeterminate or variable ranging over X, sometimes called a *free* variable. The expression "$P(x)$" is not itself an assertion but becomes one when instantiated by replacing the free variable x by a specific instance x_i in X.

The *universal quantifier* $\forall$ is a way of making a general assertion about the set or universe X. It is attached to the indeterminate x, and the result conjoined with $P(x)$. The resultant expression

$$(\forall x)P(x)$$

(in words, "for all x, or for any x, or for every x, x has the property P; $P(x)$ is true)" clearly tells us something about X itself. The expression can be *interpreted* as a string of conjunctions:

$$(\forall x)P(x) = P(x_1) \vee P(x_2) \vee \ldots \vee P(x_n) \vee \ldots.$$

This is an example of a *general proposition* on X.

Clearly, we can always infer the *particular* $P(x_i)$ from the general proposition $(\forall x)P(x)$. This is *deduction;* in this form, one can see why deduction is regarded by many as trivial and why deductive sciences like mathematics tend to be despised by such people. However, such views completely miss the point; in mathematics, for example, the whole art is bringing a situation to a point where the final deducation *is* trivial. But that art itself is the antithesis of triviality; on the contrary, it is often of the most exquisite creativity and beauty.

The problem of *induction,* on the other hand, is a complementary attempt to go the other way; it is an attempt to establish a general proposition $(\forall x)P(x)$ on the basis of particular instances $\{x_{i_1}, x_{i_2}, \ldots, x_{i_r}$ in X for which it is known that $P(x_{i_k})$ holds.

Put another way: let S be a subset (a set of *samples*) of X, and let s be a free variable that runs only over the subset S. When we can conclude, from a knowledge that

$$(\forall s)P(s),$$

we must also have

$$(\forall x)P(x)?$$

That, roughly, is the general problem of induction; the effective passage from particulars (on S) to general on X).

Obviously, without further structure, the problem of induction cannot be solved in general; that much was originally pointed out by Aristotle, if not his predecessors, and elevated by Hume into a complete rejection of empiricism. By "empiricism" here, we mean the establishment of general truth by judicious sampling, so that what happens on an appropriately chosen sample set S can in fact be extrapolated to all of X.

The further structure necessary to deal with inductive problems effectively lies, of course, in the property P with which we are dealing. If P is simply any old property, then we cannot extrapolate from any sample set S to the whole set X, and the problem is indeed unsolvable by sampling. But if P, as a property, itself manifests what we may call *contagion,* so that the truth of $P(x_i)$ itself implies the truth of $P(x_j)$ for some other x_j's in X, then the problem of induction can be dealt with. Certainly in science, and in mathematics as well, we seldom deal with entirely arbitrary properties. Continuity, for example, is a contagious property, in the precise sense that what is true *at* a point remains true *near* the point.

The most vivid mathematical example of such contagion is embodied in what is correctly called *mathematical induction.* Roughly, induction in this sense requires only two things of a property P: (1) $P(1)$; and the crux of it, (2) for an arbitrary integer n, $P(n)$ *implies* $P(n + 1)$. The conclusion is: under these conditons, $P(n)$ is true for every integer.

Mathematical induction is, in fact, all we need to generate the whole of Number Theory from the existence of the number "1", and the ability to "add 1" to any integer. This indeed is the substance of the Peano Axioms for arithmetic, in which mathematical induction is the only inferential procedure available.

This idea will turn up again and again in our subsequent discussion; as we shall see, it permeates both science and mathematics in profound and insufficiently appreciated ways. I shall have much more to say about these matters at the appropriate time; see especially the discussion of *recursion* and its consequences, starting in section 4C below.

The efficacy of induction, the demonstration of general truths by extrapolating from samples, depends on the characteristics of the *property P* we are looking at. If, as is the case with mathematical induction, the property is contagious enough, then a sample of one is sufficient. Therein lies its strength.

2E. On the Generality of Physics

As already noted, it has been the prevailing sentiment in science today, as it has been for centuries past, that physics is the general, and hence, that biology is only a particular. In the present section, I will try to clarify what this assertion means.

In some ideal sense, of course, this assertion about physics is trivially true. Ideally, the *aspiration* of physics, its dream, is to encompass material nature in all of its manifestations. Organisms, as a part of material nature, clearly fall within this compass. Hence, from such an ideal perspective, biology indeed becomes part of the physical whole.

But I am not, in fact, addressing this ultimate ideal. When I use the word "physics," I am talking about *contemporary* physics; physics as it exists now, today, embodied concretely in all the books and journals reposing in the physics sections of all the libraries in the world. It is a very different matter to suppose, as reductionism requires, that biology is a particularization or specialization of *contemporary* physics.

To assess the "level of generality" of contemporary physics, or indeed of any other scientific or mathematical discipline, in any kind of absolute terms is an extremely difficult thing. It raises in fact a metatheoretic question; a question *about* the theory, not a question *within* the theory. Intuitively, the "level of generality" of a theory characterized the class of situations with which the theory can cope, the class of phenomena it can in principle accommodate. How, if at all, can such a thing be measured?

It is instructive, in this regard, to look at the Theory of Numbers in pure mathematics, where the situation is much more under control. Number Theory has historically been plagued with conjectures (really inductions, based on limited experience or sampling with small numbers), which no one has ever been able either to prove or produce a counterexample (disprove). Is Fermat's Last Theorem a theorem? How about the Goldbach Conjecture, that every even number is the sum of two odd primes? Is Number Theory general enough, even in principle, to cope with these very specific situations?

The situation is made even more interesting as a result of Gödel's celebrated work on undecidability in Number Theory, which we shall see much more of as we proceed. In brief, Gödel showed how to represent assertions *about* Number Theory *within* Number Theory. On this basis, he was able to show that Number Theory was not finitely axiomatizable. In other words: given any finite set of axioms for Number Theory, there are always propositions that are in some sense theorems but are unprovable from those axioms (unless, of course, the axioms are inconsistent to begin with—in which case everything is a theorem). The conclusion here is that *every* finitely axiomatized system of Number Theory is too special, in some abstract, absolute sense. But there is no way of telling whether a specific assertion or conjecture about numbers is provable, or disprovable, or undecidable (unprovable) within such a system.

If this is already the situation in Number Theory, how much more complicated to ask similar questions about physics. But that is exactly the question raised by reductionism; it is an assertion, or conjecture, or belief, pertaining to the generality of contemporary physics itself. And indeed, it is not a conjecture based on any *direct* evidence (as, say, Goldbach's Conjecture in Number Theory is), but rather on indirect (circumstantial) evidence, insofar as evidence is adduced at all. In short, it rests on *faith*.

As we have seen, generality is hard to assess in absolute terms. It is not as hard to assess *relative* generality, the generality of one theory with respect to another; I gave examples of this in earlier sections. In the present context, it is instructive to compare the level of generality of contemporary physics with that of physics as it was, say 100 years ago.

Indeed, the "evolution" of physics as a science is nothing but a continual increase in its generality. A century ago, for instance, phenomena of atomic spectra, of radioactivity, of chemical bonding, and many others were outside the ken of physics. This could be seen in at least two ways: either physics provided no way to even begin to treat such problems, or else, when it did, it gave blatantly wrong answers. From our present perspective, a century later, we can see that classical mechanics and classical thermodynamics were simply unable to cope with such problems. We can see now that the problems were *conceptual* ones and that theoretical physics a century ago was too restricted, too narrow, too special, to accommodate them, even in principle. At the time, of course, such problems were given no such interpretation; they were regarded as purely *technical* matters, requiring no modification of the underlying conceptual apparatus but only more cleverness in deploying that apparatus. In fact, it was authoritatively argued at just this time that physics was essentially complete as a conceptual discipline, that the primary remaining task for physicists was merely to measure its parameters with every greater accuracy. In short, just before physics was to undergo the most profound revolutions in its history, it was widely believed that it had already achieved its ideal state.

I have already indicated above how the two primary revolutions in physics in our century, relativity and quantum mechanics, came about. In each case, nineteenth-century physics had inadvertently restricted itself by means of two apparently harmless, and indeed quite invisible, hypotheses: that the velocity of light was infinite and that the quantum of action was zero. This was, of course, tantamount to requiring all velocities to be small, all energies low; nineteenth-century physics was really only the physics of material systems satisfying those conditions. In all other situations, it was forced either to stand mute or to be egregiously mistaken.

By contrast, the physics of the twentieth century (i.e., contemporary physics for us today) can accommodate arbitrary energies, arbitrary velocities. The generalizations that comprise modern physics have enormously extended the range of phenomena we can now accommodate. In the process, we have learned that the range of nineteenth-century physics consisted of what today are *limiting cases;* situations that are atypical, and *nongeneric,* in the context of the physics of today.

How could this happen? Only because, in the nineteenth century, material situations involving high energies and high velocities were *rare.* They could neither be produced technologically nor observed (or at least recognized) with the instruments then available. Because they were rare, and for that reason only, they could be discounted as technical anomalies, or given *ad hoc* "explanations" within the then-existing conceptual framework.

But we have seen this before (see section 1A above). Physics has discounted biology because organisms are *rare* in the class of material systems. On that basis, they have been presumed *special,* nongeneric *in the class.* But that is exactly the same kind of assumption that, as we have seen, characterized nineteenth-century physics; high-velocity, high-energy phenomena were *rare* in the nineteenth century, and hence, regarded as special and nongeneric. The truth in the latter case was quite the contrary of this; it is rather low-velocity, low-energy situations that are really from our present perspective the rare, atypical, nongeneric ones, however, prominent they may appear in everyday experience.

The generalizations that have allowed contemporary physics to encompass relativistic, and quantum-theoretic, phenomena have, of course, radically altered the face of physics itself; it is a far different thing today than it was a century ago. But in a certain sense, there is still a great deal of nineteenth-century physics, and seventeenth-century physics, that has survived these revolutions and generalizations intact. What other tacit presuppositions and limitations lie lurking there?

As I noted earlier, with respect to biological phenomena, contemporary physics is in exactly the same situation that nineteenth-century physics faced in the atomic and cosmological realms: it either stands mute or it gives the wrong answers. That is the simple fact. Once again, as in all similar situations in the past, the claim is that purely technical matters are involved and that the problem is simply one of *specializing* what already exists in an appropriate way. But history shows it to be at least equally likely that the problems are not technical but conceptual, that contemporary physics remains too special to accommodate the class of material systems we call organisms.

If so, it becomes a matter of finding, and removing, whatever tacit hypotheses are limiting the generality of contemporary physics in these directions. We cannot find them by looking at contemporary physics itself;

it, and everything in it, are already the consequences of imposing these very hypotheses. Rather we must retreat to an earlier conceptual stage. In the process, we shall find in fact that contemporary physics embodies a number of such restrictive hypotheses, and we shall see in detail the dramatic effects of removing them.

Chapter 3

Some Necessary Epistemological Considerations

I FEEL IT is necessary to apologize in advance for what I am now going to discuss. No one likes to come down from the top of a tall building, from where vistas and panoramas are visible, and inspect a window-less basement. We know, intellectually, that there could be no panoramas without the basement, but emotionally, we feel no desire to look at it directly; indeed, we feel an aversion. Above all, there is no beauty; there are only dark corners and dampness and airlessness. It is sufficient to know that the building stands on it, that its supports, its pipes, and plumbing are in place and functioning.

3A. Back to Basics

Scientists, especially, are impatient with their basement of epistemology and ontology and what they call metaphysics. We are proud professionals who have reared not just a building but a temple, a monument. Our calling our genius, is in fact never to stop building but rather to push what we have received ever higher, to enlarge, to adorn, to move upward. Why take the time and the trouble to descend and contemplate anew uncongenial things that were settled long ago?

And so I must apologize for conducting the reader on a necessary trip back to the basement. It will only be for a short while, I promise, and I also promise that what we do there will be of importance.

Moreover, having been at the summit allows one to see the basement with new eyes. Many years ago, the mathematician Felix Klein wrote a series of books called *Elementary Mathematics from an Advanced Standpoint;* as the title suggests, the books constituted an illuminating return to the lower floors of mathematics, from the perspectives of the higher. And one should also ponder the reminiscence of Einstein:

> I sometimes ask myself, how did it come that I was the one to develop the Theory of Relativity? The reason, I think, is that a normal adult never stops to think about problems of space and time. These are things which he has thought of as a child. But my intellectual development was retarded, as a result of which I began to wonder about space and time only when I had already grown up. Naturally, I could go deeper into the problem than a child. . . .

Sufficient reason indeed to beat a strategic retreat. And that is what we shall do for the next several sections.

3B. The First Basic Dualism

Science is built on dualities. Indeed, every mode of discrimination creates one. But the most fundamental dualism, which all others presuppose, is of course the one a discriminator makes between self and everything else.

As Descartes argued long ago, the only absolute, undoubtable certainty lies here, and he put it with the ultimate terseness: *Cogito, ergo sum.* By *"cogito,"* Descartes meant the entire spectrum of the activity of his mind: perception, cognition, ideation, will, imagination. Also, although this gave him a great deal of trouble, because it involved something beyond this warm circle of certainty, he meant the capacity for action.

Thus, as Descartes says, we know our *selves,* without even having to look, by an immediate kind of direct apprehension and with a knowledge that brooks no skepticism.

Oddly, I have not been able to find a really good word that incorporates all of the activities of the self that we know with such immediate certainty. In physics, the word *observer* is often used, but as we shall see, this is too passive. There is the Freudian word *ego,* which is more encompassing than "observer" but has become ringed round with connotations irrelevant or misleading for our purposes. Perhaps best, as we have done, to continue to use the noncommittal word "self," though it seems rather drab and humble, and certainly insufficiently technical, for the exalted role that Descartes gives it.

At any rate, we know our self with ultimate certainty, even though this knowledge is *subjective;* it cannot be experienced as we experience it by anything else; at best it can only be reported. As noted, we encompass as belonging to the self, or contained within it, our perceptions, our thoughts, our ideas, our imaginings, our will, and the actions that spring from them. This is the *inner world.* Everything else is *outside.*

What else is there? Whatever it is, I shall call it the *ambience.* Most of us believe there are indeed many things in our ambience; this is the *external world,* the world of objective reality, the world of phenomena. That world is important to us, because our bodies are in that world, and to that extent at least, we must seriously care what goes on out there.

Much more could be, and has been, said about this fundamental dualism between the self and its ambience, but we shall need no more than the simple fact of its existence. Science, in fact, requires both; it requires an external, objective world of phenomena, and the internal, subjective world of the self, which perceives, organizes, acts, and understands. Indeed, science itself is a way (perhaps not the only way) of bringing the ambience *inside,* in an important sense, a way of importing the external world of phenomena into the internal, subjective world that we apprehend so directly. I shall have much more to say about this when we come to the idea of natural law, and especially, the idea of a *model* (see section 3F et seq. below). Indeed, as we shall see, the fact that inner, subjective models of objective phenomena *exist* connotes the most profound things about the self, about its ambience, and above all, the relations between them.

3C. The Second Basic Dualism

Our first basic dualism has separated the universe into a self and its ambience. For each of us, this separation is absolute, indubitable, and unequivocal, though it may be different for different selves. Our second basic dualism concerns the way we partition our ambiences, the way we *manage* our perceptions of the external world.

At this level, we have no universal principles to guide us, nothing *given* to us, like the distinction between the inner world of the self and the outer world, what we called the ambience. It rests rather on a consensus *imputed* to the ambience, rather than on some objective and directly perceptible property of the ambience. It is the dualism between *systems and their environments.*

Roughly speaking, a *system* in the ambience is a collection of percepts that seem to us to belong together. It would be hard to imagine a less precise definition of anything, but that is inherent in the very idea of system. The abstract concept of *systemhood* is indeed a very difficult one to grapple with, as is the related notion of *set-ness.* It is at the same time familiar in the concrete garb of everyday experience and alien when we attempt to characterize it in isolation, as a thing in itself, apart from any specific material embodiment.

Indeed, in mathematics, set-ness is such a basic and familiar notion that it took two thousand years for it to be recognized explicitly; even then, it took a strange mind (as contemporaries reported Cantor was) to see it and to deal with it. Once it was pointed out, and its central role in mathematical thought made explicit, then everyone saw it. Indeed, within a generation, and in the teeth of paradoxes it had already spawned, David Hilbert was saying, "From the *paradise* created for us by Cantor, let no man drive us forth."

The notion of system-hood is at that same level of generality and plays the same kind of role in our management of the ambience. As noted, it segregates things that "belong together" from those that do not, at least from the subjective perspective of a specific self, a specific observer. These things that belong together, and whatever else depends on them alone, are segregated into a single bag called *system;* whatever lies *outside,* like the complement of a set, constitutes *environment.*

The partition of ambience into system and environment, and even more, the imputation of that partition to the ambience itself as an inherent property thereof, is a basic though fateful step for science. For once the distinction is made, attention focuses on *system.* Systems and environment are thenceforth perceived in entirely different ways, represented and described in fundamentally different terms. To anticipate somewhat, system gets described by *states,* which are determined by observation; environment is characterized rather by its effects on system. Indeed, it is precisely at this point that, as we shall see, fundamental trouble begins to creep in; already here.

The growth of science, as a tool for dealing with the ambience, can be seen as a search for special classes of systems into which the ambience may be partitioned, such that (1) the systems in that special class are more directly apprehensible than others, and (2) everything in the ambience, any other way of partitioning it into systems, is generated by, or reducible to, what happens in that fundamental class. Newtonian mechanics, for instance, thought it had found such a class; so, today, does quantum theory. But it is, above all, a *special* class, embodying an equally special way of coping with the system-environment dualism itself. Whether this is enough is, at root, the basic question.

3D. Language

An essential part of the inner world of any self is one's language. It is a way, or reflects a way, of organizing percepts and perhaps even of generating them.

Language itself creates, or embodies, new dualism distinct from (but in many ways parallel to) those we have already discussed. Indeed, language is a unique and anomalous thing, whose acquisition, and even more, whose correct deployment, is a kind of miracle. I cannot dwell on these matters here but rather will concentrate on its essential role as an intermediary between the self and its ambience, and between one aspect or part of the self and another.

The first basic dualism inherent in language is that (1) it is a thing in itself and (2) permits, even requires, referents external to itself. These embody respectively what we will call the *syntactic* aspects of language and its *semantic* aspects. Roughly speaking, syntax pertains to what language *is,* as a thing in itself, while semantics pertain to extralinguistic referents. These referents may involve the self, or the ambience, or both, or even neither.

Let us consider syntactic aspects first. Syntax involves its own inherent dualism, which may be roughly described as the dualism between *proposition* and *production rules.* From a syntactical point of view, divorced from any external referents, propositions in the language are in general not *about* anything and are described entirely in terms of conventional symbol vehicles: letters, words, sentences, and so forth. The production rules are themselves propositions, but they do have referents, namely, other propositions *in the language.* Their role is essentially a dynamic one, to enable the construction of new propositions from given ones, or the analysis of given propositions into simpler ones.

The syntactical production rules of a language are its internal vehicles for what I shall call *inferential entailment.* The rules thus allow us to say, without consulting any external referent, that *one proposition, or group of propositions, implies others.* More generally, inferential entailment is a relation between propositions and means precisely that there is a string of production rules whose successive application will take us from some of them to the others.

Just as nobody has been able to characterize an organism in terms of a discrete list of properties, no one has been able to characterize a "natural language" (let us say English) in terms of a list of production rules. Indeed, if it were possible to do this, it would be tantamount to saying that a (natural) language can be completely characterized by syntactic properties *alone,* i.e., made independent of any semantic referents whatever. There have indeed been deadly serious attempts to do precisely this (see my remarks on formalization below). They have all failed, often rather dramatically, indicating (what might be obvious) that, in general, semantics cannot simply be replaced by more syntax. Nevertheless, the attempt to do so has served to extract various kinds of syntactical "sublanguages"; these will play an analogous rule, in the external world of the self, to the segregation of systems in one's external world or ambience. Indeed, as we shall soon see, there is more than just an analogy here.

We shall understand by a *formalism* any such "sublanguage" of a natural language, defined by syntactic qualities alone. That is, a formalism is a finite list of production rules, together with a generating family of propositions on which they can act, without any specification or consideration of extralinguistic referents. Thus, a formalism, as a fragment of natural language, *could* be "about" something (i.e., endowed with extralinguistic referents), but it *need not be.* A formalism, by its very nature, carries with it no "dictionary" associating its propositions with anything outside itself. It is propelled entirely by its own internal inferential structure, as embodied explicitly in its production rules. These and these alone determine the relations among the propositions of the formalism, which we have called inferential entailment.

As we shall see, the extraction of a formalism from a natural language has many of the properties of extracting a system from the ambience. Therefore, I shall henceforth refer to a formalism as a *formal system;* to distinguish formal systems from systems in the ambience or external world, I shall call the latter *natural systems.* The entire scientific enterprise, as I shall soon argue, is an attempt to capture natural systems within formal ones, or alternatively, to embody formal systems with external referents in such a way as to describe natural ones. That, indeed, is what is meant by *theory.*

A prominent trend, indeed a characteristic one, of contemporary science and mathematics is to try to dispense with extralinguistic referents entirely and replace them with purely syntactic structures that only recognize and manipulate the symbols of which the propositions themselves are built. This process is, naturally enough, called *formalization.* It involves the internalization of semantic referents, in the form of additional, purely symbolic, syntactic rules. It has never been better described than by S. C. Kleene (1950); I have quoted it often before, but it does no harm to repeat his words again here:

> We are now about to undertake a program which makes a mathematical theory itself the object of exact mathematical study. . . .
>
> The result of the mathematician's activities is embodied in propositions . . . we can contemplate the system of these propositions.
>
> . . . As the first step, the propositions of the theory should be arranged deductively, some of them, from which the others are logically deducible, being specified as the axioms (or postulates).
>
> This step will not be finished until all the properties of the undefined or technical terms which matter for the deduction of the theorems have been expressed by axioms. Then it should be possible to perform the deduction treating the technical terms as words *in themselves without meaning.* For to say that they have meanings necessary to the deduction of the theorems, other than what they derive from the axioms which govern them, amounts to saying that not all of their properties which matter for the deductions have been expressed by axioms. When the meanings of the technical terms are thus left out of account, we have arrived at the standpoint of formal axiomatics.

This idea of formalization, that the semantic aspects of language can *always* be effectively replaced by purely syntactic ones, will turn out to be another place where really serious trouble creeps in. Indeed, Gödel showed in effect that it was already false for Number Theory. It will turn out to be closely related to the reductionistic idea that there is always a "largest model," as I shall later describe in detail (see chapter 8). For the moment, however, I simply suggest the reader bear in mind the basic conclusion we can distill from the discussion above: *natural language is not a formalization.*

The study of formal systems is what comprises the subject of (in the broadest sense) *mathematics.* Its object is the universe of formal systems, just as real and significant a part of the self's internal world as are the natural systems one extracts from one's ambience. Seen in another way, mathematics is the study of inferential entailment, the art of extracting inferents from premises or hypotheses.

I conclude this brief consideration of language by pointing out two apects of natural language that will play key roles in what follows but that never end up as part of formalisms. These are (1) the use of the *interrogative,* to which I have already alluded, and (2) the use of the *imperative.* The latter, for example, is universally presupposed, even in mathematics; an algorithm, for example, is nothing but a strong of imperatives, ordering us to apply specific production rules to specific propositions, assuring us that *if* we do so, some definite end *will thereby be entailed.* In the world of natural systems, similar lists of imperatives constitute recipes, protocols, blueprints, and the like, which govern *fabrication.* But, as will become apparent, the entailment process *embodied* by algorithms or recipes is very different than that governing their *application.* The difference, indeed, is precisely the difference between fabrication and physiology, which I contrasted earlier (see section 1C above). And the difference between them will provide another central feature of our overall enterprise.

3E. On Entailment in Formal Systems

Entailment is perhaps the central concept in the present work, as it is in the entire scientific enterprise and beyond. We have just met with one kind of entailment, namely, inferential entailment or implication within a formalism. As I stated, it constitutes a relation between propositions in a formalism; *P entails Q,* or *P implies Q,* if there is a string of production rules that take us from the proposition P to the proposition Q in the formalism. There are many ways to say this: P is a premise and Q is a conclusion; P is a hypothesis and Q is a theorem; etc. And of course, a formal system itself can be looked at as the totality of propositions Q entailed as theorems by the "axioms," using the system's inferential machinery or production rules in this fashion.

One of the endearing features of a formal system is that, *if it is consistent,* then truth necessarily percolates hereditarily from the postulates to the most remote theorem; inference can never take us out of a class of true propositions. By *consistent,* we mean only that proposition of the form $(P \vee \bar{P})$ are never theorems, where $\bar{P}$ is the negation of P (whatever P may be). Hence if the postulates are arbitrarily called "true," that same adjective may be safely applied to every inference or entailment drawn from them. In everything that follows, we always suppose we are dealing with formal systems that are consistent in the above sense. As we saw earlier, though, consistency is something we can seldom be absolutely sure of.

In any case, suppose that we step outside our formalism and contemplate one of its theorems P. That is, let us look at, or observe, our own formalism. From that perspective, we can *interrogate;* we can ask, for instance, what is it about that system which makes P a theorem? Put another way, we can ask: why is P true in the system? What, if anything, *entails* "the truth of P"?

To this question, several distinct kinds of answers can be given.

1. Obviously, the truth of P in a formal system depends on that system's axioms. For, by definition, P is entailed from precisely those axioms; this

is exactly what makes *P* true. So the axioms, which entail *P*, also play a role in entailing "the truth of *P*."

2. A different kind of role within the system is played by the production rules, the machinery that actually does the entailing. Different production rules, with the same axioms, might not entail *P*. Therefore, "the truth of *P*" is also partially entailed by the rules of inference that govern the system.
3. The actual entailment of *P* by the axioms, which constitutes the *proof* of the theorem *P*, involves more than just the inferential rules themselves. It is an explicit *list* of these rules, to be applied sequentially in a particular order. The first rule in the list is applied to the axioms themselves. The second rule is applied to the lemma arising therefrom. The third rule is applied to the result, and so on. The last rule on the list produces *P* itself. Such a list of production rules, each in an imperative mood, constitutes an *algorithm* or *program*. The exhibition of such an algorithm or list is another way of entailing "the truth of *P*" in our system.

These ways of answering the question "why is *P* true?" are clearly different and independent. Any of them can be changed or modified, independently of the others, and any such change may modify the status of "the truth of *P*." We can change an axiom, without touching the inferential rules or any algorithm drawn therefrom. We can change an inferential rule, without changing either the axioms or the list of which that constitutes our algorithm. Finally, we may change the algorithm, without affecting either the axioms or the rules themselves. And if we do make any one of these changes, thereby changing the status of "the truth of *P*" in the modified system, there is no guarantee that we can make further changes that restore "the truth of *P*" to its original status. Thus, for instance, if we change an axiom in such a way that *P* is no longer a theorem, we may not be able to make a corresponding change in the production rules, and/or the algorithm that constituted the original proof of *P*, so as to make *P* a theorem again. This is what we mean when we say the three ways we have answered the question "why is *P* true?" are independent.

Before going further, let us also note the obvious fact that the kinds of changes we have contemplated all come from *outside the formalism*. There is obviously no mechanism *within* the formalism for changing an axiom, or a production rule, or for applying a different algorithm. Furthermore, from the standpoint of the formalism, *anything* that happens outside is accordingly *unentailed*. To put it another way: A question like "why has this axiom been replaced by another?" can have *no answer at all* within the formalism itself. This is our first glimpse of a peculiar thing, which will later become of prime importance: namely, that though formal systems allow us to talk about entailment in a coherent way, from their standpoint everything important that affects them is itself *unentailed.*

So I have given three different kinds of answers to the question "why is *P* true?" The discussion I have provided should remind the reader of something we have seen before: namely, the Aristotelian discussion of the causal categories. Indeed, we have paralleled three of his four categories of causation; specifically, if we call the theorem *P* an *effect,* we may identify his idea of material cause of *P* with the axioms of a formalism, his idea of efficient cause of *P* with its production rules, and his idea of formal cause of *P* with the specification of a particular sequence or algorithm of production rules, generating a corresponding trajectory of propositions from axioms to *P*. At the moment, this should be regarded as no more than a curiosity; our main point is simply to indicate that *the Aristotelian analysis can be applied to any entailment structure,* simply by (as he did) asking "why?" about it.

The reader may not be surprised to note that we do not see a formal analog of Aristotle's fourth causal category, which he held to be the most significant; namely, the category of final cause. Perhaps the reader would be more surprised to see finality asserted; for centuries past, it has been part of the essential core of science itself that science and finality are incompatible. I shall discuss this matter, which is in fact one of the key inheritances we receive from Newtonian mechanics, when we come to consider natural systems (see sections 4I, 5K below). But already at this point, in the internal world of formalism, we can say a few important things about finality.

In what follows, *telos* is not involved; as we shall soon see, finality and teleology are in fact very different things. In completely formal terms, we may note that final causation appears anomalous, when compared with the other categories of causation. Formally, to say that something is a final cause of *P* is *to require P itself to entail something;* in every other case, to say that something is a cause of *P* means only that it *entails P*. Final cause thus requires something of its effect *P;* in all other cases, nothing is required of *P* beyond the passive fact of its entailment.

Moreover, in addition to requiring its effect *P* to entail something, a final cause of *P* must entail *the entailment of P itself.* It is this peculiar reflexive character of final causation, visible here in purely formal terms, that is primarily responsible for its anomalous position.

Let us recast these considerations in another way, since they will become crucial for us later. In any formalism, there is a kind of natural flow from axioms to theorems, very much like the familiar unidirectional flow of time. Indeed, the formal analog of "time" is embodied in the idea of *sequence,* the order of application of production rules or inferential operations in proofs and algorithms. This flow of "formal time" is irreversible, just as real time is, and as we shall see, for exactly the same reasons. In it, the axioms are always *earlier* than any of their consequents; a proposition P is *later* than another Q if it is implied by it, if there is a proof of P with Q as hypothesis.

The three "traditional" causal categories (formal, material, and efficient causation) always respect this flow of "formal time", in the sense that "cause" Q always precedes effect P. Final causation gives the *appearance,* at any rate, of violating this flow, in the sense that the effect of P seems to be acting back on the causal process that is generating it; it appears that the "future" is actively affecting the "past." I say "appears" because this (traditional) interpretation of finality confuses P with its final cause; it is not the effect P, but the final cause of P, that must operate on the process by which P is generated. The temporal anomaly remains, however; final cause clearly cannot fit *within the same temporal sequence* in which the other causal categories harmoniously operate.

The rejection of finality in science is usually cast in this temporal context, in the form of an unspoken "Zeroth Commandment" permeating all of theoretical science: "Thou shalt not allow the future to affect the present."

The upshot of this discussion of finality is the following: in purely formal terms, a concept of final causation requires modes of entailment that are simply not generally present in formalisms. In a typical formalism, a proposition P entails only its consequents under the given inferential rules. Further, there is generally nothing in a formalism that *entails an entailment.* Both are required to make a concept of finality formally meaningful; formalisms generally contain neither. There is, however, nothing inherently impossible about them, and for the moment, I simply suggest keeping an open mind about formalisms *rich enough in entailment* to admit a meaningful category of final causation, at least in the limited sense I have described. In effect, I am suggesting, on formal grounds, the possibility of *separating finality from teleology,* of retaining the former while, if we wish, discarding the latter. Later (see chapter 5), I will argue that the incorporation of finality into our scheme of things, in the form of the additional modes of entailment it requires, is not only possible, it is crucial.

3F. On the Comparison of Formalisms

I have argued above that mathematics, in the broadcast sense, is the study of formalisms and that formalisms, in their turn, are parts of natural language whose inferential or entailment structures are defined in purely syntactic terms. But mathematics comprises more than this. For one thing, mathematics selects, from the plethora of formalisms available, only a rather small number of protracted scrutiny; the selection process, which gives mathematics its form as a human activity, is not itself part of any formalism. For another thing, once a variety of formalisms is so selected, the formalism themselves become elements in a potential mathematical universe.

More precisely, formalisms may be *compared,* in terms of their respective inferential or entailment structures. When are two formalisms, which *look* different in terms of their axioms and production rules really "the same formalism" in terms of the body of propositions (theorems) they generate? When does one formalism subsume another, so that the second can be in some sense generated from the first, or embedded in it? And above all, is the machinery for dealing with such questions, i.e., with the comparison of formalisms, itself a formalism? In the present section, I shall briefly describe some of these questions, which will arise again in other contexts later; it turns out to be instructive to look at them here, in a purely formal context, free of the murky epistemological embroidery that obscures the fundamental issues elsewhere.

The problem of comparing formalisms is at root one of classification or taxonomy; it does for mathematics what Linnaeus did for biology. The basic underlying idea for such comparisons goes back to Euclid and is embodied in the idea of *similarity* or *congruence;* related geometric figures can be brought into coincidence by applying some kind of transformation to the underlying space. How related the figures are is measured in some sense by the complication of the transformation required to make them coincide. For instance, Euclidean congruence requires only rigid motions (rotations, translations, reflections); Euclidean similarity (equal angles) allows transformations of a more *general* character, and so on until we get to topology (in which figures are called congruent if they is simply a continuous map that brings them into coincidence).

This kind of taxonomic classification permeates mathematics (and much more beyond); the general study of similarity analysis in fact involves one of the archetypal aspects of human thought. It would take us too far afield to consider it in detail here (see *AS*), but I offer one more example from mathematics itself.

Let S be a linear vector space, a set of things called vectors. These vectors may be added and may be multiplied by scalars (numbers), subject to the normal rules governing these operations, which everyone knows. Inserting a coordinate system into S allows us to express any vector as an array of numbers and converts vector addition and scalar multiplication to the familiar manipulation of these arrays. Different coordinate systems will clearly associate each vector with a different array of numbers.

A *linear transformation* T of one such vector space S_1 into another S_2 is an ordinary mapping of the set of vectors S_1 into the set of vectors S_2, which respects the vector operations; it maps sums to sums, and scalar multiples to scalar multiples, in the familiar way. As such, it is already an instrument for *comparing* S_1 and S_2, but that is not our present concern.

If S_1 and S_2 come along with a particular coordinate system, then T itself translates into a particular array of numbers, a *matrix.* The same transformation T will give rise to different arrays, different matrices, in different coordinate systems. Thus, just looking at matrices, i.e., at arrays of numbers, we can ask when two such arrays come from the same linear transformation T, only seen in different coordinate systems. This clearly establishes a relation on *matrices* (similarity); two matrices are similar if they "differ by coordinate transformations." Specifically, if A and B are matrices, then they are similar if and only if there is a coordinate transformation g_1 of S_1, and a coordinate transformation g_2 of S_2 such that, for every vector v in S_1

$$g_2 \mathrm{A}(v) = B g_1(v)$$

or, in more familiar abstract terms, if

$$g_2 A \, g_1^{-1} = B.$$

This kind of relation can be embodied in a diagram that exhibits the mappings involved, namely,

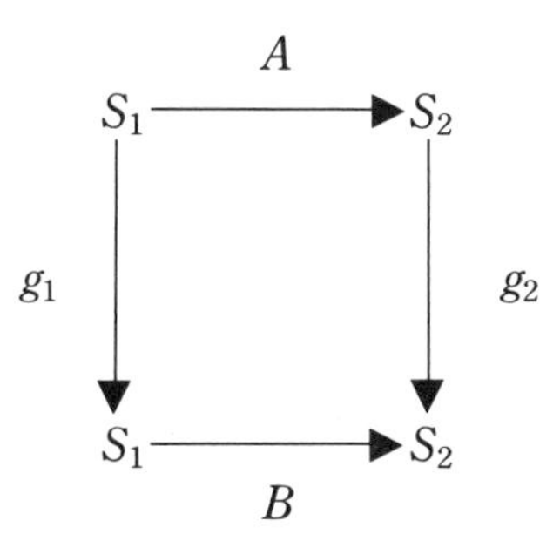

The similarity condition requires this diagram of mappings to *commute;* the two paths going from upper left to lower right in the diagram must always agree.

I have taken this digression, first, to show that mathematical objects besides geometrical figures can be compared (i.e., congruence is an idea of more general currency than just geometry), and second, to exhibit in a familiar situation the kind of archetypal commutative diagram that is the hallmark of comparison in mathematics.

Now let us return to the man question, namely, the comparison of *formalisms*. What there is to be compared here, i.e., to be brought into congruence, to the extent possible, is the inferential structures that characterize the formalisms. The result, if we are successful, will be a commutative diagram of the type we have just seen.

So let us say we have a *formalism* F_1, and another formalism F_2 we wish to compare it with. Each of these formalisms possesses, of course, its own inferential structure, its own set of production rules and axiomatic propositions on which they act to generate the consequents or theorems that constitute the respective systems. I will indicate these autonomous inferential structures in the schematic way in figure 3F.1.

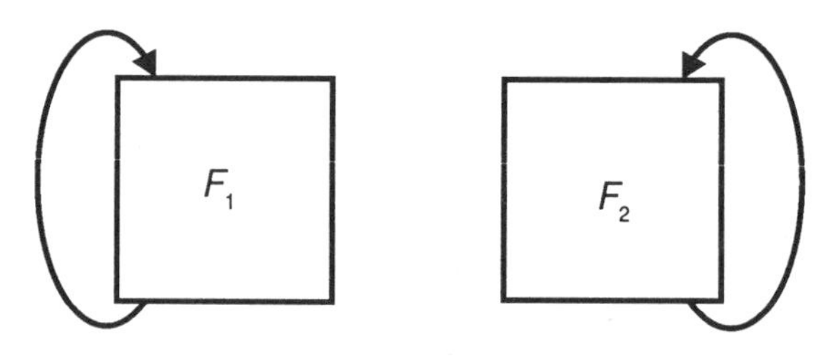

FIGURE 3F.1

So far, F_1 and F_2 are only, so to speak, talking to themselves. In order to compare them, we need to get them to talk to each other, or better, to express what each formalism says to itself in the language of the other. In other words, as with comparing any two languages, we need to make a *dictionary,* in fact, a pair of dictionaries. The first of them, which will translate from F_1 to F_2, I shall call the *encoding dictionary,* or simply the *encoding* (of F_1 into F_2). The other, translating from F_2 back to F_1, I shall call the *decoding dictionary,* or just *decoding.* I note explicitly that, in general, these dictionaries are not just inverses of each other; we do not require any relation at all between them.

So far, these encodings and decodings are just set-theoretic maps from propositions in one system to propsoitions in the other, which mandate a *synonymy*. Putting all the arrows together, we have a diagram (figure 3F.2) of the form which is just figure 3F.1 augmented by the dictionaries.

So far, the dictionaries are arbitrary. As such, there is no reason why they should respect the inferential structures in F_1 and F_2. More explicitly, given a proposition P in F_1 the diagram gives us two ways to deal with it:

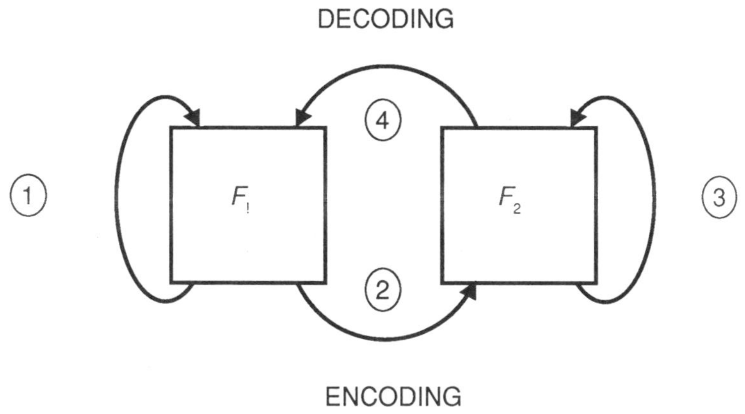

DECODING

1 F_1 4 2 F_2 3

ENCODING

FIGURE 3F.2

1. We can apply to P the inferential machinery in F_1 itself and look at the resultant propositions entailed thereby.
2. We can *encode* P (if it is indeed allowed by the encoding dictionary we are using) into a proposition P' in F_2. Then, using the inferential machinery of F_2, we can look at the propositions in F_2 entailed by P'. And finally (again, if it is allowed by the decoding dictionary), we can decode these back into propositions in F_1.

Using general dictionaries for encoding and decoding (i.e., using an unconstrained specifications of synonymy), there is no reason why these two procedures should agree or coincide. That is, using the numberings of figure 3F.2, we will generally find that

$$1 \neq 2 + 3 + 4$$

Thus, the diagram *does not commute.* In such a case, our choice of encoding and decoding establishes no relation between the inferential structures of F_1 and F_2.

It may, however, be that we can *find* encodings and decodings for which the diagram of figure 3F.2 *always commutes.* in such a case, we have in fact brought at least a part of the inferential machinery of F_1 into congruence with a corresponding part of the inferential machinery of F_2. We will then say that F_2 is a *model* of F_1, or equivalently, that F_1 is a *realization* of F_2. We will also say that a *modeling relation* exists between the two inferential structures.

We should note parenthetically here that the word *model* is overworked and has been used in a whole host of different, sometimes unrelated, and even contradictory senses. There is, for example, a well-developed *Theory of Models,* employed mainly to study the consistency of axiom systems in Foundational Studies in mathematics. The use of the term "model" in this context is not quite the same as mine; in fact, it is more closely akin to what I have called "realization." However, since I will never use the word "model" in any other sense than the one I have specified, there will be no danger of *internal* confusion. I simply want to warn the reader at this point that the danger of equivocation on "model" is unusually great.

A host of truly marvelous things follows from the establishment of a modeling relation between formalisms (see *AS).* I shall describe some of these later, in a more appropriate context. The main thing about it, of course, is that we can use the inferential structure of the model to study that of its realization, to *predict* in effect, from an encoded hypothesis P' (via the pathway $2+3+4$ in the diagram) theorems of F_1 from theorems in F_2.

An enormous amount of this kind of formal modeling of one kind of inferential structure in another occurs in mathematics. Indeed, whole fields (algebraic topology, for example, see section 5I below) consist of nothing else. I have in *AS* described and illustrated these activities at great length and there is no need to repeat it here. I shall rather content myself with pointing out some of its general features, which will be of importance to us later.

The first matter of importance is to note that, from the standpoint of the formalisms being compared, *the encoding and decoding arrows in figure 3F.2 are unentailed.* In fact, they belong to neither formalism, and hence, *cannot* be entailed by anything in the formalisms. The comparison of two inferential structures, like F_1 and F_2, thus inherently involves something outside the formalisms, in effect, a *creative act,* resulting in a new kind of formal object, namely, the modeling relation itself. It involves *art.*

The second matter concerns whether this creative act can itself be formalized, i.e., whether the study of comparison of formalisms is itself a formalism. In a nutshell, the answer is *yes, in a sense.* The name of that formalism is the *Theory of Categories;* the qualification is that Category Theory, like Number Theory, like Set Theory, or like natural languages themselves, cannot be *formalized,* in the sense of Kleene quoted previously. Indeed, many mathematicians have wondered aloud, over the years, whether Category Theory is even a part of mathematics.

However, Category Theory comprises in fact the general theory of formal modeling, the comparison of different modes of inferential or entailment structures. Moreover, it is a stratified or hierarchical structure, without limit. The lowest level, which is familiarly understood by Category

Theory, is, as I have said, a comparison of different kinds of entailment in different formalisms. The next level is, roughly, the comparison of comparisons. The next level is the comparison of these, and so on.

The final matter I wish to draw attention to here is the following. In a precise sense, any formalism F has a "biggest" formal model, namely, F itself. But as we have seen, even in mathematics we find entailment structures that are not formalizable. Let us use natural language for the sake of illustration. There is certainly a lot of entailment in natural language; suppose we want to model it. That is, suppose we want to place a natural language into the box occupied by F_1 in figure 3F.2 above. Then what?

In fact, we can find a host, an *unlimited number, of distinct formalisms* F_2 that we can put into a modeling relation with the language. No one of them, nor indeed, no aggregate of them, can replace the language, in the sense of completely duplicating its entailment structure. In short, the totality of formal models of something that is not itself a formalism to begin with is

1. indefinitely large, and
2. is not itself a formalism.

These will turn out to be pregnant and profound conclusions. We can get some sense of their ultimate import if we replace the word "formalism" with the word "machine." Much lies ahead of us before we can make this kind of substitution sensible. But it will turn out that the conclusions we have drawn above, which have so far concerned only those formal entailment structures specifiable in terms of syntax alone, can be exported to comparison of *any* kind of entailment structures whatever. I now turn to the question of whether there are any others, and if so, what they might be like.

3G. Entailment in the Ambience: Causality

I now turn from the internal, formal world of the self to the external world that constitutes its ambience, the world we have come to look upon as populated by *natural systems* and their environments. I thus turn to the world of science, in the broadest sense.

The fundamental question for us, at this point, is the following: is there, in this external world, any kind of *entailment,* analogous to the inferential entailment we have seen between propositions in a language or formalism? Obviously, if there is not, we can all go home; science is not only impossible but also inconceivable.

This kind of question has always been difficult, because we come by our knowledge of the ambience at second hand. As philosophers have pointed out for millennia, all we perceive directly are our selves, together with sensations and impressions that we normally interpret as coming from "outside" (i.e., from the ambience), and that we merely *impute,* as properties and predicates, to things in that ambience. The things themselves, the *noumena,* as Kant calls them, are inherently unknowable except through the perceptions they elicit in us; what we observe are *phenomena,* which are to an equally unknowable extent corrupted by our perceptual apparatus itself (which of course also sits partly in the ambience).

We can simplify things somewhat if we ask the more restricted question: is there any kind of entailment at the level of phenomena? Or, stated otherwise: does it *appear* to us that a phenomenon can entail another? The problem is still difficult, because entailment at this level is a *relation* between phenomena (just as inferential entailment is a relation between propositions), and we usually do not directly perceive relations. Indeed, a relation between phenomena depends on a double imputation: the first from sensation to phenomena, the second from phenomena to relations between them. Thus, if our knowledge of phenomena is already once removed from the ambience, any talk of entailment, or any other kind of relation between phenomena, is twice removed. On top of all of this is a further problem, that what we *do* perceive is only a sample of what we *could* perceive and the problems of induction arising thereform; see section 2C above.

It goes without saying that most of us can adduce the most compelling, convincing subjective evidence for believing that, and acting as if, there are indeed entailment relations between phenomena. But the question is rife for rampant skepticism; despite the combined efforts of countless philosophers, there is no way to *entail* the existence of such relations from anything else (i.e., from anything in the internal world of the self, or anything that the self draws from, or imputes to, the ambience). To such a skeptic, indeed, there is little to distinguish science from paranoia (which is basically a search for, or a belief in, entailments that are in some sense not there).

Nevertheless, it is hard to believe, for instance, that we could use natural language, in its semantic role of bringing external referents inside, if there were not a great many phenomenal entailments; semantic language by its very nature imputes hordes of entailments to the ambience, without going really dramatically astray. For this, and similar (albeit subjective) reasons, we will suppose that relations of entailment do indeed exist between phenomena; the question then becomes not whether, but when, such relations hold.

It was, of course, Aristotle who associated the notion of entailment between phenomena with the question "why?" and answered it with a "because." Indeed, the pair consisting of the question "why *A*?" and the answer "because *B*" precisely asserts an entailment of *A* by *B*, and hence, an *explanation* of *B* in terms of *A*. In this way, entailment relations between phenomena are subsumed under the general framework of *causality*. To the extent that science is the study of entailment relations between phenomena, Aristotle correctly identified science with the study of "the why of things" and scientific explanation with the elucidation of causal sequences.

Historically, Aristotle elaborated his view of the causal categories in terms of human artifacts (i.e., statues, goblets, houses) rather than in terms of animate or inanimate nature or in terms of formalisms. Nevertheless, as we have seen, his analysis holds good wherever there are relations of entailment *of any kind,* even in the world of formal systems, where entailment means inference. Accordingly, his analysis also applies to the world of natural systems that populate the ambience; as we shall see abundantly later, it permeates the whole of contemporary science, though in such a shrunken and distorted form that it takes a special effort of retrieval to make it manifest.

We shall thus accept this view, that entailment relations can exist between phenomena and that their study comprises causality; hence science and causality are to that extent synonymous.

I turn now to the last of our preliminary considerations, namely, the establishment of relations *between* the two entirely different kinds of entailment we have been considering. I have talked about *inferential entailment* in internally generated formalisms, governed by inferential rules that generate new propositions from given ones. And I have talked about *causal entailment,* relating phenomena arising in the ambience or external world. My final task is to show that these two entirely different modes of entailment are themselves related. The assertion of this relation is embodied in the concept of Natural Law; the crucial instrument in establishing the relation is the concept of *model.*

3H. The Modeling Relation and Natural Law

Over the preceding several sections, we have been mainly concerned with the concept of entailment. We have in fact found two entirely different realms in which entailment is meaningful. First, in the interior world of the self, I have called attention to inferential entailment and embodied it in the inferential structure of formalisms or formal systems. Second, in the outer world of the ambience, I identified a different kind of entailment, entailment between phenomena. As I said, this is the province of causality.

There are many parallels between these realms of entailment, which I am, in fact, presently in the process of making explicit. We have already seen, for example, that the Aristotelian analysis that led to his ideas about causation in the ambience actually apply to any realm possessing a notion of entailment; thus, in particular, we saw (see section 3E above) that his analysis was equally meaningful for syntactic entailment in formal systems. We also saw, in the preceding section, that the modes of entailment manifested by different formal systems may be compared; the vehicle for such comparison of entailment in different formal systems was the establishment of a modeling relation between them.

In the present section, we shall see that the concept of a model provides in fact a general method for comparing entailment structures of *any* kind. Just as the Aristotelian analysis may be applied to any mode of entailment, so too can modeling relations be established, *mutatis mutandis,* between entailment structures of arbitrary kinds. I shall now indicate in particular how we can compare inferential entailment in a formal system with causal entailments, relating a bundle of phenomena that we extract from our ambience and identify as a natural system.

A modeling relation between causal entailment in a natural system and syntactic entailment in a formal one provides a concrete embodiment of the concept of *Natural Law.* It is worth spending a moment discussing Natural Law, for it provides the explicit underpinning on which all of science rests.

Natural Law makes two separate assertions about the self and its ambience:

1. The succession of events or phenomena that we perceive in the ambience is not entirely arbitrary or whimsical; there are relations (e.g., causal relations) manifest in the world of phenomena.
2. The relations between phenomena that we have just posited are, at least in part, capable of being perceived and grasped by the human mind, i.e., by the cognitive self.

Science depends in equal parts on these two separate prongs of Natural Law. The first, which says something about the ambience, asserts that it is in some sense orderly enough to manifest relations or laws. Clearly, if this is not so, there can be no science, also no natural language, and most likely, no sanity either. So it is, for most of us at any rate, not too great an exercise of faith to believe this.

The second part of Natural Law says something about ourselves. It asserts that the orderliness of the ambience is (to some unspecified extent)

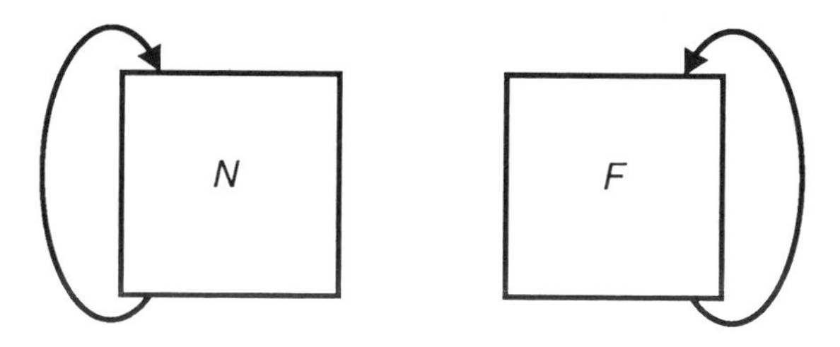

FIGURE 3H.1

discernible to, and even more, is articulable by, the self. It asserts then that the posited orderliness in the ambience can be matched by, or put into correspondence with, some equivalent orderliness within the self.

In other words, the first part of Natural Law is what permits science to exist in the abstract. The second part of Natural Law is what allows scientists to exist. Clearly, concrete science requires both.

I am now going to show how the modeling relation provides an explicit embodiment of Natural Law. Specifically, the causal entailments manifested by a natural system provide the orderliness required of the ambience. Inferential entailment in a formal system is a way of providing the orderliness required of the self. The art of bringing the two into correspondence, through the establishment of a definite modeling relation between them, is tha articulation of the former within the latter; it is in effect science itself.

My discussion will exactly parallel the one given earlier. Thus, let us suppose we are given a natural system N, and a formal system F (figure 3H.1). Just as before, the two arrows schematically represent the respective entailment structures; inference in the formalism F, causality in N. And as before, a comparison of these different kinds of entailment structures requires the establishment of *dictionaries,* one for encoding the phenomena of N into the propositions of F and another for decoding from propositions of F back to phenomena in N. Such dictionaries give us exactly the same diagram we have seen before (figure 3H.2).

The concept of such a dictionary is, of course, precisely what endows a syntactically defined formalism with external referents. In this light, the commutativity of a diagram like that of figure 3H.2, and the resultant modeling relation it embodies, is just an expression of the possibility of using *syntactic* truth and *semantic* truth consistently. As I have said several times before, natural language as we know it could not otherwise exist.

Intuitively, the encoding arrow ② is associated with the notion of *measurement.* In physics, for example, a measurement process is precisely geared to associate a *number* with an event or phenomenon in N. A number is an abstract object, a mathematical entity. Thus, a meter serves basically

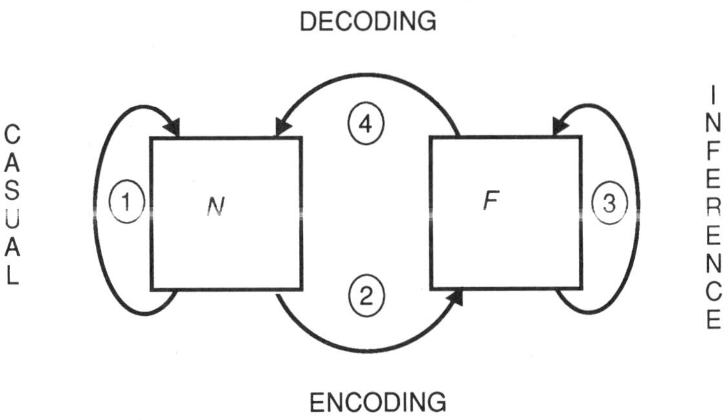

DECODING

CASUAL

INFERENCE

ENCODING

FIGURE 3H.2

as a transducer, associating numbers with phenomena. It is in fact one of the basic beliefs of physics, made quite explicit in quantum theory, that every observation, i.e., every material interaction of the self with its ambience, can be equivalently expressed in terms of an appropriate family of numerical measurements, but that is not really important to us now. Indeed, it may well be false; it does not change the argument.

It is not perhaps generally appreciated, especially by experimentalists (i.e., by those who actually perform measurements) that any measurement, however comprehensive, is an act of *abstraction,* an act of replacing the thing measured (e.g., the natural system N) by a limited set of numbers. Indeed, there can be no greater act of abstraction than the collapsing of a phenomenon in N down to a single number, the result of a single measurement. From this standpoint, it is ironic indeed that a mere observer regards oneself as being in direct contact with reality and that it is "theoretical science" alone that deals with abstractions.

In any event, the decoding arrow ④ in figure 3H.2 represents a *de-abstraction,* the association of a phenomenon in N with a proposition in F. It is thus a kind of "inverse" measurement, going from propositions to events.

Just as before, we note that there are two separate paths in the diagram: namely,

① and ② + ③ + ④.

Each of these paths takes us from phenomena in N to phenomena in N. The first of them (the path ①) represents causal entailment within N; it is essentially what an observer, who simply sits and watches what happens in

N, will see. The second path, however, involves more. First we must encode, via the arrow ②, from phenomena in *N* to propositions in *F*. Next, we must use these propositions as *hypotheses*, on which the inferential machinery of the formal system *F* may operate. That machinery, the entailment structure within *F*, is what we have denoted by the arrow ③; it generates theorems in *F*, entailed precisely by the encoded hypotheses. The final step is, to the extent permitted, to *decode* these theorems back to the phenomena of *N*, via the arrow ④. At this point, the theorems we have thus generated become *predictions* about *N*.

The formal system *F* is then called a *model* of the natural system *N* if we always get the same answer, whether we follow path ①, or whether we follow the path ②+③+④ As before, the establishment of a modeling relation between *N* and *F* serves to bring their respective entailment structures into at least a partial coincidence. To that extent, then, we can learn about one by looking at the other. And to that extent, modeling relations are nothing more than embodiments, in concrete situations, of natural law as I discussed it above.

As I have described it, the modeling process compares causal entailment in *N* with inferential entailment in *F;* if we are successful in establishing such a relation, then *F* is the model; *N* is a *realization* of that model. But it is essential to note that the roles of *N* and *F* can be interchanged. That is, instead of starting with a natural system *N*, and looking in effect for a formalism *F* that models it, we could start with a formal system *F* and ask for a natural system *N* whose causal entailment provides a model for inferential entailment in *F*. This, it will be recognized, is not simply an interchange of the arrows ② and ④ in figure 3H.2 above. It constitutes what I shall call *the realization problem*. At this level, it appears innocent enough; as we shall see, however, its consideration involves modes of entailment falling completely outside contemporary science. Indeed, the entailments required to deal with it are closely related to those that characterize *finality*, as I have described above. Ultimately, it will turn out that the absence of precisely such modes of entailment from our inherited scientific arsenal is what makes biology, and especially the origin of life, so hard.

We can already see some peculiar epistemological issues inherent in the diagram of figure 3H.2 that did not arise earlier. Specifically: what is the status of the encoding and decoding arrows in that diagram? We already saw, in the exactly similar diagram of figure 3F.2, that the encoding and decoding arrows were themselves *unentailed*. But at least they could themselves be considered as formal objects, since at that point we were comparing syntactic entailment in two formalisms. But now, we are com-

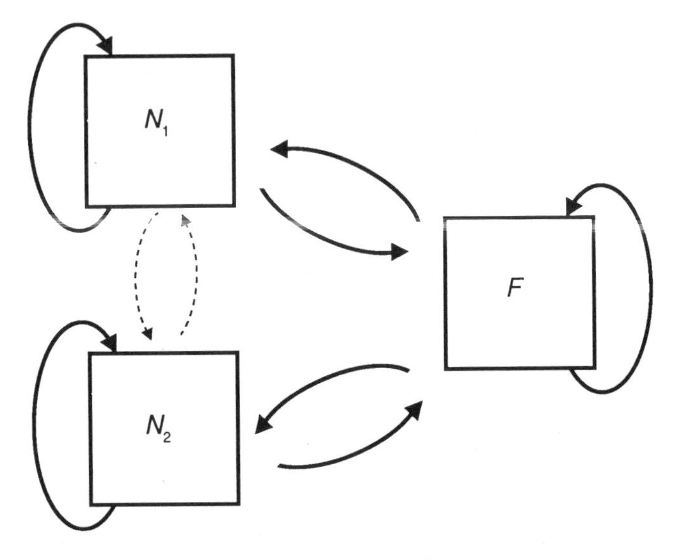

FIGURE 3H.3

paring syntactic entailment in a formalism with causal entailment in a natural system. The encoding and decoding arrows in this case are still *unentailed,* but it is no longer clear *how* they could be entailed, or from what. These arrows are not part of the natural system N, nor even of its environment; although they pertain to the ambience, they do not belong to it. Neither do they belong to the formal world of the self either; they look like mappings, but they do not compare formal objects; hence they cannot be mappings in any formal sense. Thus these arrows, which play the central role in comparing causal and inferential entailment, and hence, in the operation of Natural Law itself, turn out to possess a new and ambiguous status, equally within, and outside of, both the self and its ambience.

Modeling relations between natural systems and formal systems are wondrous things in many ways. Indeed, the innocuous-appearing diagram of figure 3H.2 has many remarkable ramifications. I have discussed many of these at great length elsewhere (see *AS*) and thus need not repeat most of it here. I can, however, use the concept of a formal model to complete my discussion of the comparison of entailment structures by showing how it can be used to compare causal entailments between two *natural* systems, say N_1 and N_2. We can do this by contemplating the diagram shown in figure 3H.3.

In this case, the two natural systems N_1, N_2 *realize a common formal system* F. It is clear that we can use the respective encoding and decoding arrows from N_1 and N_2 to F to construct the dotted arrows in the diagram,

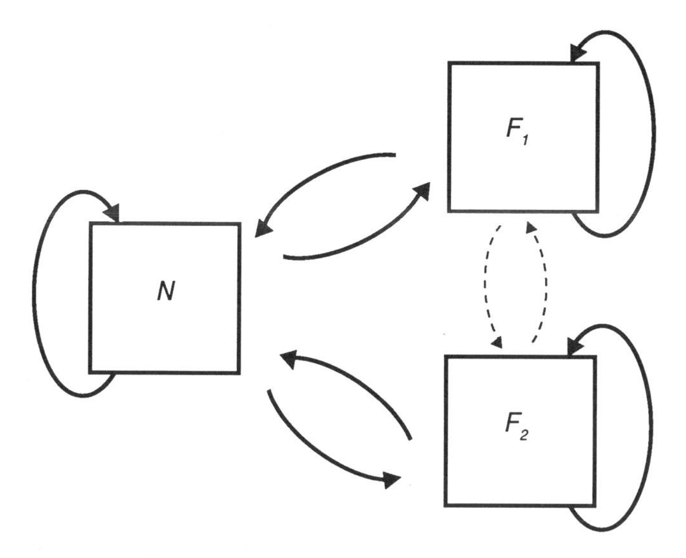

FIGURE 3H.4

in effect, to construct a dictionary from phenomena in N_1 to phenomena in N_2. The commutativities required by the assertion of modeling relations between N_1 and F, and N_2 and F, ensure that the whole diagram commutes, in the obvious sense. Hence N_1 and N_2 are *models of each other;* their respective structures of causal entailment are, to that extent, brought into congruence through the intermediary of a common formalism F, whch they separately realize.

This modeling relation between two natural systems N_1, N_2 is of the most profound importance; I shall call it *analogy.* It means that we can learn about one natural system N_1 by looking at a different, perhaps completely different, natural system N_2. This is familiar enough; under the rubric of *scale modeling,* engineers have long exploited the relation of analogy. In physics, we have already (see section 2C above) alluded to the "Mechano-Optical Analogy" of Hamilton and the wonders it spawned. Still more dramatic, perhaps, are the vistas opened by analogy for the further elucidation of causal entailment in natural systems that are, in conventional physical terms, of entirely different structures: organisms and "machines", organisms and social systems. This is another way of seeing, what I alluded to earlier, that reduction to a common set of material constituents is not the only way, nor even a good way, of comparing natural systems.

The complementary diagram to that of figure 3H.3 is exhibited in figure 3H.4. Roughly, it is the situation in which a single natural system N possesses two formal models.

Once again, we can construct the dotted arrows in the diagram from the posited modeling relations between N and F_1, and N and F_2, and thus compare syntactic entailment in F_1 and F_2. Clearly, the *formal* relationships between F_1 and F_2 arising in this way bear immediately on the problem of reductionism. For instance, we can ask, for *the class of all formalisms F that model N,* what is the formal structure of such a class? Is there a largest model in the class? Or are there causal entailment structures (i.e., natural systems N) in the ambience, as there are within the self, for which there is no largest model, for which the class of all formal models does not determine a formal model?

Thus we have come full circle, back to the problems with which we started. And with this, we are nearly at the end of our tour of the epistemological basement of contemporary science. In a moment we shall turn to a contemplation of what can be built thereon.

31. Metaphor

As we have seen, the modeling relation is intimately tied up with the notion of prediction. Natural Law, as embodied in modeling relations, thus equips us to look into the future of things; insofar as the future is entailed by the present, and insofar as the entailment structure itself is captured in a congruent model, we can actually, in a sense, pull the future of our natural system into the present. The benevolence of Natural Law lies in assuring us that such miracles are open to us, but it does not extend to telling us how to accomplish them; it is for us to discover the keys, the encodings and decodings, by which they can be brought to pass.

Thus, if we want to predict, to tell the future, there is much work to be done in the present. But most of us are lazy; we would like to decouple benefits from costs, to the fullest extent possible. A large part of the cost imposed by Natural Law, in return for the benefit of prediction, lies in finding the right encodings. But to what extent do we really need these encodings? Perhaps we can presume a little on Natural Law and get away without them.

At first sight, this does not look feasible; if we want to predict something about a particular natural system, it seems evident that we must specify that system, i.e., that we must encode it somehow. But on the other hand, prediction itself involves nothing more than *decoding* from a model or formalism. Certainly, if we already have a model, we can forget about how our natural system was encoded and obtain correct predictions just by decoding from the formalism. Perhaps this situation has wider currency;

perhaps we can decode something about the future of some natural system without explicitly encoding its present at all.

If we can do this, then Natural Law provides us to that extent with a nonspecific crystal ball whereby we seem to derive its benefits with only half the work. This is the essense of *metaphor:* decoding without encoding, in a sense, only the top half of our modeling relation.

All science, and biology in particular, is replete with such metaphors. They have in fact been of profound importance in the history of science and in many areas continue to play a major role; in fact, they constitute what there is of theory in these areas. Let us accordingly look briefly at a few of them.

Perhaps the most important for our purposes is the machine metaphor of Descartes, to which I have already alluded (see section 1C above), and about which I shall have much more to say (see chapter 7 et seq.). It asserts that things about machines can be decoded into predictions about organisms, without the benefit of any specific encodings going the other way. Machines thus become our crystal ball, our one-way mirror for looking into the organic world, without needing to look out again. We do not need to dwell further on the crucial role this metaphor continues to play in shaping the outlook of biology.

Another one of enormous current importance, which looks different from the Cartesian metaphor, but which is in fact closely related, is what may be called the open system metaphor. This is of relatively recent origin, at least in its present form; it was most explicitly articulated by Ludwig von Bertalanffy in the mid-1930s. Bertalanffy drew attention in particular to the metaphorical relation between what happens in the vicinity of stable point attractors (stable steady states) of open systems and the empirical facts of embryonic development: pattern generation or morphogenesis. In this metaphor, we seek to decode from the former into the details of the latter, again without the benefit of any specific encodings going the other way. It was this general metaphor, embodied in particular submetaphors by Rashevsky and Turing, that sent physicists like Prigogine scrambling to modify thermodynamics to accommodate them; see *AS*. In a somewhat different direction lie the more formal metaphors of Catastrophe Theory, first proposed by Thom. There are nowadays many variants of this basic metaphor, all being pursued with great diligence.

At root, such metaphors are pursued in the belief, or expectation, that they can in fact be turned into models. Thus, for example, the expectation is that any particular natural system, like a developing frog embryo, can be explicitly encoded into *some* formal open system, in such a way that a modeling relation is thereby established, or, more generally, that any organism can be explicitly encoded into some machine in such a way as to complete the modeling diagram so that commutativity holds. Meanwhile, the metaphor itself allows us to derive many of the benefits of the modeling relation even in the absence of these encodings.

To proceed metaphorically in the above sense is, of course, not an unreasonable thing to do. It is also clear, however, why experimentalists find such metaphors troubling and why they occupy an anomalous position in what passes nowadays for philosophy of science. For by giving up encoding, we also give up *verifiability* in any precise sense. Thus, experimentalists interested specifically in, say, a developing sea urchin, derive no tangible help from a metaphor. They need something to verify, couched in terms of some specific observation, or experiment, that they can perform. That is to say, they need precisely what is missing in the metaphor; they need the encodings. Hence the general indifference, if not active hostility, manifested by empiricists to theory couched in metaphorical terms. Metaphor is indeed immune to such verification; insofar as science is identified with verification, as it is currently fashionable to do, metaphor is not even science. Nevertheless, it is clear that metaphor can embody a great deal of truth. And as with all crystal balls, it does have the irresistible attraction of offering something for free.

Metaphor exists on the purely formal side as well. In the Theory of Categories, for example, it manifests itself in the concept of functor (see section 5I below). The turning of metaphor into model, in those terms, is expressed in the concept of natural transformation. The whole idea of the Theory of Categories arose initially in this way, and it is illuminating to continue to regard it in this light. In some sense, it is precisely the unique metaphorical aspects of Category Theory that generate qualms in many mathematicians regarding it, which run quite parallel to those of any empiricist.

Chapter 4

The Concept of State

As WE have seen, the ambience or external world is traditionally regarded as being composed of systems (natural systems) and their environments. In the present section, we will begin to look into this matter more deeply, using the concepts we have already developed.

4A. Systems and States

Central to the notion of natural system is the attendant notion of *state*. As I remarked earlier, and as we shall see in more detail in the present chapter, *system* and *state* have become essentially coextensive; systems are described in terms of their possible states, while their environments are not (and indeed, cannot be).

As is true with all the deep concepts of contemporary science, the idea expressing systems in terms of states, and everything that happens in systems in terms of state transitions, goes back to Newtonian mechanics. In effect, Newton did for science what formalization did, or tried to do, for mathematics some three centuries later. There is indeed a profound parallel between Newtonian particle mechanics and the pure syntax of formalizations; in each case, everything is supposed to be generated from structureless, meaningless elements (particles in one case, symbols in the other), pushed around according to definite rules (forces in one case, production rules in the other). In each case, all that ultimately matters is the spatial disposition of these elements, their *configuration*. The concept of state does for particle mechanics what proposition does for formalism; it expresses a "meaningful" configuration of basic elements on which the syntactic rules can act. The sequence of state transitions of a system of particles, governed by Newton's laws of motion, starting from some given initial configuration, is then the analog of a *theorem* in a formalism, generated from an initial proposition (hypothesis) under the influence of the production rules. And just as formalization in mathematics believed that everything could be formalized *without loss,* so that all truth could be recaptured in terms of syntax alone, so particle mechanics came to believe that every material behavior could be, and should be, and indeed must be, reduced to purely syntactical sequences of configurations in an underlying system of particles.

Hence the power of the belief in reductionism, the scientific equivalent of the formalist faith in syntax. Though of course Newtonian mechanics has had to be supplemented and generalized repeatedly, the basic faith in syntax has not changed; indeed, it has been bolstered and made more credible by these very improvements. And there has as yet been no Gödel in physics to challenge that credibility directly. But there *is* biology.

The syntactical ideas, first and most potently manifested in Newtonian particle mechanics, have had another ramification; namely, the *form* taken by mechanical description (i.e., a set of states, on which are superimposed a set of rules governing change of state) has become the universal currency for describing systems of any kind. Thus, although in most cases we may not be able to get our hands on the underlying particles, and the rules governing them, we still exclusively utilize the Newtonian *language;* whether the field be chemical kinetics, or population dynamics, or economics, or power engineering, we still try to define an appropriate set of states, and a set of dynamical laws, in terms of which the behaviors of interest of a system can be generated and understood. From a purely reductionistic viewpoint, such an approach is only a stopgap, a formalism not yet fully formalized, but while we are waiting for the ultimate reduction to be effected, we can still use the form of the reductionistic language, if not yet its substance. And in fact, this is exactly what we do. However, when we divorce the language of mechanics from its susbtance, we run into certain problems. In Newtonian mechanics we have the luxury of being very explicit about what *state* means and what its properties are. When we leave the substance of mechanics behind, the concept becomes murky, and it becomes increasingly easy to equivocate on it. Nevertheless, it remains basic to every way we presently have of dealing with material reality; indeed, it is generally taken for granted as the starting point for every mode of precise scientific investigation. Hence it is appropriate that we begin our own analysis here.

4B. Chronicles

I will begin by stepping back a bit, by supposing that we do not yet have a notion of state at our disposal. In effect, I will retreat to the level of percepts and perceptions and treat the self as a pure observer. The idea of state, being a concept and not a percept, thus does not yet enter the picture at all. Thus, all we have is the self looking out at its ambience. What does it see?

All the self *can* see is a sequence of percepts, ordered by its subjective sense of time. We suppose that the self can choose *which* percepts it will look at (in more sophisticated language, which variables it will measure) and whether it will look continuously or sample at discrete intervals. (Subjective) time is itself a complicated concept (see *AS*), but it is a primitive that we can take for granted at this level. Thus, the result of the self looking at its ambience is only a tabulation; a list of what is seen, indexed by when it is seen. Such a list we shall call a *chronicle.*

Chronicles can thus be completely arbitrary things, at least insofar as *what* is tabulated in them is concerned. Weather bureaus, stock exchanges, census takers, and a host of other familiar institutions provide endless streams of them. In the scientific realm, they are *data.* To the historian, entirely concerned with what happened when, they are the very stuff of existence.

To the (applied) mathematician, and to the statistician, a chronicle is simply a *time series.* It is thus a way of associating events, or attributes of events, with numbers (instants of time). It is even an *effective* way of associating events with numbers; all we need to do is wait for the appropriate instant to occur and then tabulate the corresponding event. In formal terms, then, a chronicle or time series is simply a *mapping* from numbers to events or their attributes; for simplicity, we can even suppose it to be a completely formal mapping from numbers to numbers.

But the self is not merely an observer. Doubtless the self has heard of Natural Law, and hence, does not believe in sequences of events that are entirely arbitrary. Further, the self may be impatient and unwilling to wait for the unfolding of time to reveal events to come; it would like to extend its chronicle into the future *before* the sequence of events does so for it. Likewise, in addition to being impatient, it may be curious about what its time series was like before the self actually started looking at it or tabulating it.

So, along with any time series (which we shall think of henceforth as a piece of a mapping from numbers to numbers) comes the urge to extend it into the future and into the past, to extrapolate it, to predict and to postdict.

The most elementary thing we can do in these directions is attempt extrapolation on the basis of the fragment at our disposal, our data. At this level, we do not know or care what the individual entries in our tabulation *mean;* we try to use the data as the basis for extrapolation into the future or into the past. This is, of course, itself a very syntactic approach; it presupposes that what we need inheres somehow in the very structure of the list or chronicle itself, apart from all other considerations and all other chronicles. Put another way, we seek to extract from the structure of the list itself something that will already *entail* those entries that are yet to come or those that have come before.

Thus, for instance, if the self is a statistician, it will look for correlations, which it may or may not find; but in any case, all it *can* find this way are properties *of the list,* and not in general of what the list represents. Clearly, Natural Law does not operate at this level. It is not a law that in general favors statisticians, though this has not inhibited their activities.

Indeed, the enterprise of trying to find the operation of Natural Law from the contemplation of arbitrary chronicles or lists of data is precisely the dilemma of experimental science itself. At root, of course, the problem is one of induction, which I have already briefly discussed above (see section 2D); it is a problem of extrapolating from a sample of a universe to the entire universe from which the sample was drawn. In the present case, we are sampling in too many ways; we are obviously sampling over an extremely limited time frame, but we are also sampling what we do observe and tabulate from the universe of what we could observe and tabulate. As always, these sampling processes corrupt us in two ways; they *lose* information precisely because they are samples, and they also *add* irrelevant information (noise), which pertains to the sampling process itself.

As we have seen, the problem of induction is generally hopeless, because arbitrary properties, simply by virtue of being arbitrary, do not reveal themselves in samples. Stated otherwise, no sample *entails* anything about a nonsampled instance. Hence, the problem of extrapolating arbitrary time series is likewise hopeless.

There are two strategies we can adopt to cope with this hopeless situation. The first of these is to retain the idea of sampling but simply sample more and different attributes. Each of these will, of course, just give us other, new time series, more data. The hope here is that multiple series will do what one alone generally cannot, namely, *entail,* on the basis of the internal structure of the *set* of chronicles, what unsampled entries must be (and especially, of course, particularly those entries that are *yet* unsampled).

The other strategy is to be more judicious in the attributes we are sampling. Although the *arbitrary* induction problem is hopeless, there are those properties that do admit sampling and extrapolation. As we saw earlier, these are properties for which entailment already exists between the *entries* in the sample or chronicle.

The concept of *state* embodies both of these strategies, more chronicles, more judiciously chosen. We will consider what is involved in the subsequent sections.

4C. Recursive Chronicles

We have seen that a chronicle may be regarded as a time series, a way of labeling events, or their attributes, by the time of their occurrence. For simplicity, in this section, we will contemplate such things in a purely formal, mathematical setting. Specifically, we will regard a chronicle as a fragment of a function or mapping

$$f: Z \to Z$$

from integers to integers.

The independent variable of the function f is to be thought of as the *instant* at which something occurs. Thus, if n is an instant, the *value* $f(n)$ is the occurrence itself (or better, some numerically measurable attribute of the occurrence). The chronicle corresponding to f is merely the list of pairs $\{n, f(n)\}$; in which each occurrence $f(n)$ is *labeled* by the instant n at which it occurs. The *label* is thus the independent variable. In practice, if an event has already occurred, and is thus tabulated in our chronicle, then to *evaluate* f at an instant n means to look down our list until we come to n and read off the corresponding value. The problem, as we have seen, is to try to extend our list or chronicle of tabulated instances, both to the left (into the past) and to the right (into the future) on the basis of what we have already tabulated.

In simplest mathematical terms, we might seek a *formula* expressing the value $f(n)$ explicitly in terms of the instant n at which it occurs, or did occur, or will occur. For instance,

$$f(n) = n^n + 5n^2$$

is a formula for some particular function f. Clearly, given such a formula, we can produce an entire chronicle, extended as far back into the past or future as we wish to go.

There are, however, several troubles with this. The worst is that, in general, we cannot determine a formula unambiguously from any sample, even if there is one. For instance, suppose I give you the chronicle

$$n \quad = 1,\ 2,\ 3,\ 4,\ \ldots,\ r$$
$$f(n) = 1,\ 4,\ 9,\ 16,\ldots,\ r^2.$$

and ask for the next number $f(r+1)$ in the sequence. Surely you will be tempted to *induce* that $f(n) = n^2$ is the formula governing the chronicle, and hence you will *predict* that

$$f(r + 1) = (r + 1)^2.$$

However, there are an infinite number of different formulas that will all produce the given chronicle; for instance

$$f(r) = r^2 + (n - r)!g(r).$$

Here, $g(r)$ is an entirely arbitrary function, and the factorial $k!$ is defined as $k(k - 1)(k - 2) \ldots 3.2.1$; hence $(n - r)! = 0\ n < r$, and nonzero otherwise. Accordingly, the "next number $f(r + 1)$" in the sequence can be anything at all. In other words, we simply *cannot entail* that next number from those that have preceded it, however long our list may be; a list thus *can never entail a formula.*

Though perhaps initially disconcerting, the absolute inability to entail a formula from a list or chronicle is inherent in our actual situation. For in the present case, the independent variable (instants) is simply a set of labels, unconnected causally or in any other way with the values $f(n)$ that they label. To have a formula for f amounts to saying that an instant, an arbitrary thing, entails via the formula the occurrence $f(n)$ with which it is associated. We cannot expect this kind of entailment; indeed, it would be disastrous if we had it. Thus, we really do not want such a formula after all.

What we do want is something very different. We want entailment between the *values* $f(n)$, and in fact, we want entailment that does not involve the independent variable n at all. In short, we want what we earlier called *contagion.*

It is in fact now easy to exhibit the sort of thing we do want. Specifically, suppose we take a mapping

$$T : Z \rightarrow Z$$

(we will interpret it in a moment) and *define* from it a function f as follows:

$$
\begin{aligned}
f(0) &= \textit{arbitrary}, \text{ say } r; \\
f(1) &= T(r) \\
f(2) &= T^2(r) = T(T(r)) \\
&\vdots \\
f(n) &= T^n(r) \\
&\vdots
\end{aligned}
$$

We say that the function f, defined in this manner, is *recursively defined;* its successive values are obtained, not by evaluating it at the successive numbers in its *domain,* but by applying a fixed operation or mapping T to its *preceding value.* In this case, if we want to think of the argument of f as "time," then an "instant" at which f is evaluated is simply the exponent of the mapping T; the "number of times" we have successively applied T to the arbitrary initially specified value $f(0) = r$.

Thus, in this situation, the function f we have defined creates precisely the kind of chronicle we are looking for. Specifically, every *value* $f(n)$ entails the next *value* $f(n + 1)$. Furthermore, the instant n is completely stripped of any intrinsic significance, precisely *because* the initial choice $f(0)$ was arbitrary.

This apparently trivial situation is the germ on which the state concept, and hence, contemporary theoretical science itself, rests.

Before embarking on the extensive justification of this last sweeping assertion, I might point out that there are more complicated situations that in a formal sense also involve recursion, entailment between chronicle entries. Perhaps the most familiar is embodied in the well-known Fibonacci sequences; in these, the first *two* members of the sequence are arbitrary, and all successive elements of the sequence are entailed from these via rules of the form

$$
\begin{aligned}
&a_1, a_2 \text{ arbitrary}; \\
&a_3 = T(a_1, a_2), \\
&a_4 = T(a_2, a_3), \\
&\quad \text{etc.}
\end{aligned}
$$

However, such modes of entailment (which, heuristically speaking, involve "time lags") are not suitable for encoding a concept of state. See, e.g., *AS*.

4D. Recursion: Some General Features

In the preceding section, we saw that recursive chronicles, generated by a mapping T and its iterates, are precisely the ones that can be induced from. In the present section, we shall explore some of the more elementary properties of such chronicles and the mappings that generate them; we shall also look at a few examples.

First, we saw that the choice of initial value $f(0)$ is arbitrary. Thus, each choice of an element from the domain of T gives us a different function f, and hence, a different chronicle.

The nature of the chronicles generated from a mapping T depends on the structure of T, as embodied in the set ϑ^+ consisting of T and its iterates;

$$\vartheta^+ = \{I, T, T^2, T^3, \ldots, T^n, \ldots\}$$

where $I = T^o$ is the identity mapping; $I(r) = r$ for every r. This set ϑ^+ is obviously more than just a set; in fact, it is a semigroup with identity, under the operation of composition of mappings.

If T happens to be $1-1$, (i.e., if distinct elements have distinct images under T), then T can be inverted. That is, there is a unique transformation T^{-1}, which undoes or reverses the application of T;

$$T^{-1}T(r) = TT^{-1}(r) = r$$

for every element r on which T is defined. It is evident that the iterates of T^{-1}, which we can write as

$$\vartheta^- = \{I, T^{-1}, T^{-2}, \ldots, T^{-n}, \ldots\}$$

then also generate recursive chronicles, which can be thought of as extrapolating a given chronicle back into the (indefinite) past.

Thus, putting ϑ^+ and ϑ^- together, to form the set of all positive and negative iterates of T, we obtain the set

$$\vartheta = \{\ldots, T^{-n}, \ldots, T^{-1}, I, T, \ldots, T^n, \ldots\}$$

which is clearly a *group* under the operation of composition of mappings. It is a special kind of group, called a *one-parameter group*. Indeed, ϑ itself can be thought of as a chronicle, as a list of things (mappings) labeled by integers (the *parameter,* or exponent n). Indeed, since we always have the nice property that

$$T^{m}T^{n} = T^{m+n},$$

we see that the labeling of iterates of T with their integral exponents is actually a *group homomorphism* between the composition operation on ϑ and addition of integers.

If we are given an integer r on which an invertible mapping T is defined, then we can generate from r the *full chronicle*

$$\vartheta(r) = \{T^{n}(r)\},$$

where n ranges over all integral values, positive and negative. We will call this the *trajectory* through r. This is in fact a special case of something we have seen before, but it is worth repeating again here: whenever we have a group of automorphisms, like ϑ, acting on a set, we obtain for free an associated equivalence relation. Specificially, two elements are called equivalent if and only if there is a group element that carries one to the other. An equivalence class under this relation is generally called a *group orbit.* In our case, equivalence of two integers r, r' means that there is a number m such that $r' = T^{m}r$, and hence, what we have called a trajectory is simply an orbit of the one-parameter group ϑ.

Accordingly, we automatically have the so-called *unique trajectory property;* two distinct trajectories cannot intersect. This follows immediately from the fact that trajectories are equivalence classes. Stated otherwise, if two trajectories intersect at one point, they are everywhere identical. This in turn means that, under the conditions we have stated, every element r on which T operates extrapolates back to a unique past or history and extrapolates forward to a unique future. Hence no two recursive chronicles, generated by the same mapping T, can have a common entry unless they are in fact the same chronicle.

Most of these nice properties disappear if T is not invertible. In that case, T must identify distinct elements of its domain; it cannot in some sense discriminate between these elements, and thus applying T irreversibly "loses information" about the past. We can still define the relation between elements r, r', which we did before (namely, the relation $r' = T^{m}r$ for some exponent m, now necessarily positive). But this relation is not an equivalence relation any more (the reader can verify that symmetry fails). In this situation, we can ony *predict;* we can uniquely extrapolate a recursive chronicle indefinitely into the future, but we cannot *post-dict,* or extrapolate it into the past. This kind of situation is therefore inherently *irreversible.* I have discussed this situation in *FM.*

Let us see what these ideas mean in the context of a few examples. Perhaps the simplest, and also the most important, example of recursion in the context of numbers is the *successor function,* whose formula is

$$\sigma(n) = n + 1$$

and which is generated by the recursion relation

$$\sigma(n+1) = T\sigma(n) = \sigma(n) + 1.$$

It will be recalled that the existence of this function was specially postulated in the Peano Axioms for arithmetic. In our context, we can think of σ as pushing us from one label or index in a chronicle to the next; its application is thus a formal analog of the "flow of time."

We will now show that recursiveness serves to pull this flow of time from the temporal labels associated with events to the events themselves, in a particularly nice way. Accordingly, let T be a mapping (it does no harm to suppose it invertible) and let f be one of the functions it generates; we recall that this means

$$f(0) = r,$$
$$f(n) = T^n(r),$$

and hence

$$f(n+1) = Tf(n).$$

Now

$$n + 1 \equiv \sigma(n)$$

for every n. We can thus write a kind of conjugacy relation between T and σ, of the form

$$f\sigma = Tf,$$

which in turn can be expressed as the *commutativity of the diagram.*

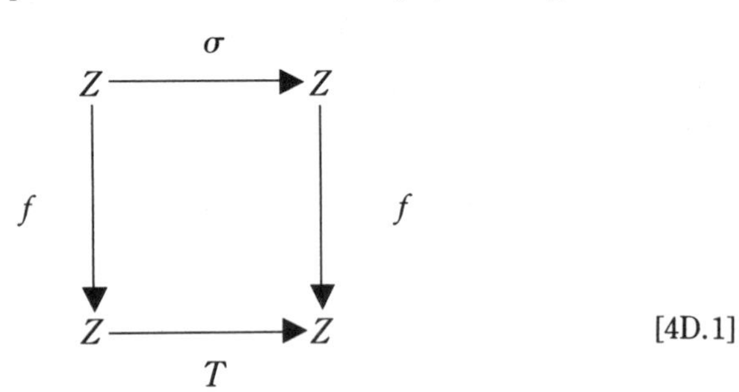

[4D.1]

Indeed, the recursiveness of a mapping $f : Z \rightarrow Z$ means precisely that we can find a T that makes the diagram commute.

Indeed, it is very easy to write down the necessary and sufficient condition for the existence of such a T. Namely, if r_1, r_2 are distinct arguments of f, such that

$$f(r_1) = f(r_2),$$

then we must also have

$$f(r_1 + 1) = f(r_2 + 1). \qquad [4D.2]$$

It is evident that a function f can be given an equivalent recursive definition *if and only if* this periodicity condition holds.

I shall conclude this section with a word about the "causal" aspects of recursive chronicles. As always, these "causal" aspects are associated with a particular interrogative: why is the nth entry is a recursive chronicle given the particular value $f(n)$? As usual, there are three different answers we can give to this question:

1. *Because* the initial entry (the zeroth entry, or initial condition) is the value r
2. *Because* of the mapping T, which generates the chronicle
3. *Because* of the exponent n, for which $f(n) = T^n(r)$

Each of these answers is associated with one of the Aristotelian causal categories; the initial condition with *material cause* of the effect $f(n)$, the mapping T with *efficient cause*, and the exponent n with *formal cause*.

In general, whenever we isolate something, either in the ambience or in the internal world of the self, and ask "why?" about it, we are treating it as an effect and inquiring about its causes. One way to cope with such questions is precisely to produce a recursive chronicle, a *history*, which starts from some convenient initial condition and takes us to the effect we inquired about through a chain of successive entailments (i.e., a trajectory) arising from a mapping T and its iterates. In the circle of ideas we are in the process of developing, which will culminate in the concept of *state*, causality manifests itself *only* through a sequence of state transitions, entailing an effect that is again a state. Although, as we shall see, a number of important tactical details intervene, this is the basic picture that permeates all of contemporary science. If there is something wrong with the picture, or especially if there is something missing from it, then the root of the trouble lies already here.

4E. On Taylor's Theorem

We have seen, in the preceding few sections, that recursive chronicles possess a host of desirable and highly suggestive properties. On the other hand, the unfortunate fact is that most chronicles are not recursive. Indeed, if our perception of the external world provides a fair or unbiased sample of that world, we are most unlikely to encounter recursive chronicles at all. On the other hand, Natural Law suggests that recursiveness must play a key role in the scientific enterprise. How can these facts be reconciled? Only through supposing that arbitrary chronicles can be embedded in, or otherwise related to, recursive ones. Thus, as will be seen, leads to our second strategy, namely, seeing if the juxtaposition of more chronicles can do more for us than a single one alone.

Let us begin by observing that the heart of recursion is the conversion of the present to the future, or the entailment of the future by the present. A chronicle of any kind is, by its nature, a *diachronic* object. Hence recursiveness is precisely a device for turning *synchrony* (what happens at a single instant) into *diachrony* (what happens over an extended series of instants).

One familiar formal device for doing precisely this has long been known to the mathematician, under the guise of Taylor's Theorem. Let us see what this theorem actually connotes; the results may be surprising.

Taylor's Theorem is a device, indeed almost a magical device, for expressing what happens *near* a point in terms of what happens *at* the point. If we have a single independent variable, and if we should perchance think of this variable as time, then Taylor's theorem in particular tells us what *will* happen, at least in the immediate future, in terms of *what is now happening.* Thus, when it applies, it is precisely a device for converting synchronic to diachronic.

More specifically, everyone learns in elementary calculus that, given a real-valued function $f(t)$ of a single real variable t, we can write

$$f(t_o+h)=f(t_o)+hf'(t_o)+(h^2/2!)f''(t_o)+\ldots,$$

where, of course, $f'(t_o)$, $f''(t_o)$, . . . are the successive derivatives of $f(t)$, *evaluated at* t_o, provided at least that the necessary successive derivatives exist. We see explicitly from this how Taylor's Theorem works in the traditional setting: we can evaluate our function f "near" any point, entirely in terms of its behavior *at* the point, and of "how near" we are to the point (the number h). We pay for this: the computation itself takes the form of an infinite series to be summed and all that entails. On the other hand, it is standard practice to *truncate* the series, to break it off after a finite number of terms (if we are lucky, the first two will suffice). Such truncation amounts to replacing the given function f by an approximation; i.e., by a *model* of f, and is hence a good example of modeling within mathematics. In any case, there are reliable estimates of the errors we make in computing the value of $f(t_o+h)$ by truncating the Taylor Series, and these are also familiar from elementary calculus.

If now we think of $f(t)$ as a chronicle (i.e., as a list of pairs $(t, f(t))$, then Taylor's Theorem asserts something very like *recursivity* of this chronicle. It is not quite recursiveness in the sense we have employed the term above, because *it depends on other chronicles;* specifically, the chronicles embodied in $f'(t)$, $f''(t)$, . . . ; the successive derivatives of f. On the other hand, if these chronicles are indeed at our disposal (and in fact only a single entry from each of these chronicles, namely, the ones at t_o, will suffice), *then $f(t)$ becomes recursive.*

At a deep level, then, Taylor's Theorem provides a glimpse at how we can make a chronicle, which in itself is not recursive, act as if it were by embedding it in a larger set of chronicles. That larger set, in the case of Taylor's Theorem, consists of a *velocity* chronicle (one that tabulates or evaluates the temporal derivatives of f), an *acceleration* chronicle (which tabulates the temporal derivative of velocity), and so on.

In this formal setting, it is easy to see that Taylor's Theorem, which extracts the future from the present, i.e., converts synchronic into diachronic, and renders very general kinds of chronicles recursive, involves no magic. Indeed, information regarding both past *and future* is embedded in the notion of derivative itself. A derivative is a two-sided limit, which in mathematical terms exists at a point only if arbitrary approximating expression of the form

$$1/h(f(t_o+h))-f(t_o))$$

converge to a common limit *from the past (i.e., $h<0$) and from the future (i.e., $h>0$)*. It is this hidden encryption of the future that is actually retrieved and made explicit in Taylor's Theorem.

As we shall soon see, since we compile chronicles in only one direction (from past to present), it involves some deep hypotheses to identify one chronicle $g(t)$ with the velocity $f'(t)$ of another. This kind of problem does not generally arise in mathematics, where a function f is given globally, in the round as it were, and its derivatives (if they exist) automatically come along with it. However, if this can be done, some rather remarkable conclusions emerge from Taylor's Theorem.

Let us begin with a simple example. We retreat from the normal habitat of Taylor's Theorem in analysis, back to the world of arithmetic and numbers. Let us take a very simple chronicle, namely

$$f(n) = (n(n+1))/2 = n^2/2 + n/2 \qquad [4E.1]$$

The reader may recognize this as the formula for expressing the sum of the first n integers. It is in fact a recursive chronicle; $f(n)$ entails $f(n+1)$. We are now going to see how we may express this entailment, in a surprising way.

Let us introduce the following new chronicles:

$$f'(n) = n + 1/2,$$
$$f''(n) = 1,$$
$$f^{(k)}(n) = 0, \text{ all } k > 2$$

and let us observe that

$$f(n+1) = f(n) + f'(n) + 1/2f''(n). \qquad [4E.2]$$

This expresses the recursion directly, in terms of $f(n)$ and the new chronicles we have introduced. But this looks, of course, very much like Taylor's Theorem; indeed, it *is* Taylor's Theorem, albeit in an unfamiliar setting.

Let us look at another example. Another very simple chronicle is given by

$$f(n) = 2^n. \qquad [4E.3]$$

It is obviously recursive; in fact,

$$f(n+1) = 2f(n). \qquad [4E.4]$$

Once again, let us introduce auxiliary chronicles, this time an infinite number of them:

$$f'(n) = 2^n(1n2),$$
$$f''(n) = 2^n(1/2!(1n2)^2) \qquad [4E.5]$$
$$\vdots$$
$$f^k(n) = 2^n(1/k!(1n2)^k)$$
$$\vdots$$

(these chronicles, it will be noted, do not take their values in integers, but that does not matter now). Putting them all together, Taylor's Theorem would suggest that

$$f(n+1) = 2^n(1 + 1n2 + 1/2!(1n2)^2 + \ldots$$

and indeed,

$$\begin{aligned} &= 2^n e^{1n2} \\ &= 2^{n+1} \\ &= 2f(n). \end{aligned}$$

These examples, and a host of others, suggest that there is indeed a deep relation between Taylor's Theorem and recursion in general. Let us now articulate what that is.

4F. Recursion and Constraints

Let us recapitulate the results of the last section. We were focusing on *recursive* chronicles, for which there exists entailment between the *values* $f(n)$ in the form of an operation T such that

$$Tf(n) = f(n+1)$$

for all values of the argument n. We then turned to Taylor's Theorem, which in its familiar guise, expresses just such an entailment rule, *with the help of other chronicles* (in this case, the chronicles we called $f'(n)$, $f''(n)$, . . .); the "temporal derivatives" of $f(n)$). If we look at the specific examples (e.g., at [4E.2] and [4E.5] above, we see that Taylor's Theorem serves precisely to give a recursion rule for $f(n)$ in the form of a "differential equation," of which $f(n)$ is a solution. Thus, for instance, the chronicle

$$f(n) = n(n+1)/2$$

is a "solution" of the "differential equation"

$$T(f) = f + f' + 1/2f''$$

This, it will be noted, is a relation among *chronicles;* it says that, for each value of the argument or independent variable n, we have

$$\begin{aligned} T(f(n)) &= f(n+1) \\ &= f(n) + f'(n) + 1/2f''(n). \end{aligned}$$

I repeat that this relation serves precisely to convert *synchronic* information (namely, what is happening at the instant n) into *diachronic* information (specifically, what is happening at another instant, $n+1$).

Likewise, we saw that the chronicle

$$f(n)=2^n$$

is a "solution" of the "differential equation"

$$T(f) = \sum_{k=0}^{\infty} 1/k!\, f^{(k)}$$

(which is of infinite order as it stands). Further, f is also a solution of *any* of the "differential equations" of the form

$$f^{(k+1)}=2/k+1(1n2)f^{(k)}.$$

which, it will be noted, are *purely synchronic,* i.e., refer *only* to a single instant n.

Thus, Taylor's Theorem provides us with a way of generating recursion rules for a very general class of chronicles. But there are costs, and one of them is the requirement for new chronicles (namely, $f'(n)$, $f''(n)$, and all the higher temporal derivatives). We shall also see, however, that each recursion rule defined in this way is associated with *identical relationships* that must be satisfied by the chronicles $\{f^{(k)}\}$ and that are embodied precisely in these "differential equations." For reasons to be made clearer in the next section, such identical relationships among the $\{f^{(k)}\}$, and hence, among the set of values $\{f^{(k)}(n)\}$ for every instant n, will be called *constraints.*

The reader will perhaps have noted that our discussion of Taylor's theorem has subtly shifted the emphasis of our discussion away from individual chronicles f to the properties of certain *sets* of chronicles, such as $\{f', f'', \ldots, f^{(i)}, \ldots\}$. In fact, our discussion of recursion of individual chronicles was primarily intended to illustrate and to motivate a generalization to such sets. All of the concepts we have introduced for an individual chronicle can indeed be extended to appropriate sets of chronicles. From now on, our discussion of recursion will shift more and more from the individual chronicle f to sets of chronicles associated with it.

Accordingly, let us begin by introducing some convenient terminology, motivated by this discussion.

We will call a chronicle f *relatively recursive* if there is a set

$$G_f = \{g_1, g_2, \ldots, g_i, \ldots\}$$

(possibly infinite) of other chronicles g_i, which allows us to write a recursion rule for f of the form

$$Tf(n) = f(n+1) = T(f(n), g_1(n), \ldots, g_i(n), \ldots)$$

valid for every n. Clearly, if Taylor's Theorem holds, then putting $g_i = f^{(i)}$ says precisely that f is relatively recursive.

Actually, Taylor's Theorem says more than just the relative recursiveness of f. To see this, we must generalize the notion of recursion from a single chronicle f to a *set* of chronicles. Accordingly, we shall say that a set of chronicles

$$G = \{g_1, g_2, \ldots, g^i, \ldots\}$$

(again, possibly infinite) is *recursive* if for each index i, there is a recursion rule of the form

$$\begin{aligned} T_i g_i(n) &= g_i(n+1) \\ &= T_i(g_1(n), g_2(n), \ldots). \end{aligned}$$

That is, the set of chronicles G is recursive if and only if each g_i in the set of relatively recursive, relative to the set $G - \{g_i\}$.

Thus, if Taylor's Theorem holds, it actually asserts that *the set of chronicles*

$$f = \{f, f', f'', \ldots, f^{(i)}, \ldots\}$$

is a recursive set.

I should perhaps warn the reader that the term "recursive set" is used in a quite different sense in the literature. That usage always refers, however, to *domains* of individual recursive functions. My usage refers to sets of *functions* that collectively admit a recursion rule, i.e., for which present values of individual functions in the set collectively entail their subsequent values. There should thus be no danger of confusion between the two usages.

I should also note explicitly that the notions of relative recursiveness of individual chronicles, and the recursiveness of sets of chronicles, say nothing about entailments between *contemporary* or synchronic values $\{g_i(n)\}$ of any of the chronicles we are dealing with. The *only* entailment asserted at the moment is embodied in the recursion rule T, which entails subsequent values from present values. It will, however, soon turn out that such contemporary entailments do exist; these are in fact precisely the *constraints* mentioned above.

As far as a single chronicle f is concerned, I assert that the following conditions are equivalent:

1. f is recursive; $f(n)$ entails $f(n+1)$ for every n.
2. The set of chronicles $f, f', f'', \ldots, f^{(i)}, \ldots$ is a recursive set.
3. Taylor's Theorem holds:

$$Tf(n) = f(n+1) = \sum_{k=0}^{\infty} 1/k!\, f^{(k)}(n)$$

4. There exist identical relations (constraints) that can be expressed in the form

$$f^{(i)} = \psi(f, f', f'', \ldots, f^{(i-1)})$$

valid at every argument n.

I repeat that the condition (3) means that the chronicle f is a solution of a certain kind of "differential equation," which is of a *diachronic* character; it relates the values of f and its "derivatives' at different arguments (i.e., different instants). The condition (4), on the other hand, says that the chronicle f is also the solution of one or more "differential equations" of a *synchronic* character, relating the values of f and its "derivatives" at a *single* argument. These are precisely the constraint relations.

4G. Coping with Nonrecursiveness: Recursion and Constraint in Sets of Chronicles

The discussion above sums up the situation in case we are dealing with a single chronicle f, which is already recursive. But it is clear that recursiveness is an unusual, nongeneric property of arbitrary chronicles. So the question arises: what do we do with chronicles that are nonrecursive?

Intuitively, nonrecursiveness of a chronicle means an absence of entailment between the values of the chronicle. If we could identify "absence of entailment" with *"lack of information,"* then it may be that the requisite "information" lies in the properties of *other chronicles,* and hence, that, by judiciously combining the values of f with those of other chronicles, we may find a more general kind of recursion rule that will entail for us the values $f(n+1)$ in terms of $f(n)$, *modulo* the new information we require.

In what follows, we will explore these possibilities. It will turn out that what is required involves relatively straightforward generalizations of the concepts developed in the preceding sections, generalizing from *individual* chronicles to *sets* of chronicles.

Before undertaking the detailed discussion, we can set out a number of possible strategies for dealing with a nonrecursive chronicle f. Perhaps the simplest (in the sense of requiring the smallest amount of additional entailment) is to try to find other chronicles $g_1, g_2, \ldots, g_i, \ldots$ such that $f(n+1)$ is entailed, for every argument n, by the values

$$f(n), g_1(n), \ldots, g_i(n), \ldots.$$

Thus, f becomes *relatively recursive,* relative to the values of the chronicles g_i. The g_i thus play the role that was earlier played by the "derivatives" $f^{(i)}(n)$ when f was already recursive. I note explicitly here that no *synchronic* entailment is involved here, nor any *diachronic* entailment among the values of the chronicles g_i. All we require here is a recursion relation of the form

$$f(n+1) = T(f(n), g_1(n), \ldots, g_i(n), \ldots).$$

One special way to do this is the following. Let us note that, quite generally, if we are given a family $\{g_1, g_2, \ldots, g_r\}$ of chronicles (suppose the family now finite, for simplicity), we can construct from them a new chronicle as follows. Let $\Phi(n_1, n_2, \ldots, n_r)$ be *any* function of r arguments, and define

$$\Phi(n) = \Phi(g_1(n), g_2(n), \ldots, g_r(n)).$$

Then Φ is a new chronicle, which we shall call an *observable* relative to $(g_1, \ldots, g_r)$; the reason for this terminology will become clear in subsequent sections. In any case, the values of Φ are determined at any instant n by the values of the g_i *at that same instant;* the relation between Φ and the g_i is purely *synchronic* as it stands. But if we now further suppose that the f_i constitute a recursive set, then by definition there is a *diachronic* entailment between the subsequent values $g_i(n+1)$ and the present values $g_i(n)$. In this situation, it is evident that $\Phi(n+1)$ is likewise thereby entailed by the $g_i(n)$ as well.
chronicle *recursible* if it can be expressed as an observable of a recursive set.

And finally, the strongest thing we could try to do with a nonrecursive chronicle f is to look for other chronicles, $g_1, g_2, \ldots, g_i, \ldots$ such that

$$\{f, g_1, g_2, \ldots, g_i, \ldots\}$$

is a recursive set.

These different ways of dealing with nonrecursive chronicles f all involve the embedding of f into a larger set of chronicles, such that the resulting *set* has some kind of recursive property. The reader will note that different kinds of entailments are involved in these distinct strategies. And I should also add explicitly that, in implementing these strategies, we always look, of course, for a set of chronicles $\{g_i\}$, which is in some sense *minimal;* i.e., we look for the *smallest* set $\{g_i\}$ that will endow our originally nonrecursive f with some kind of relative recursive character.

As I will soon indicate, each of these more general modes of recursion, or diachronic entailment, is also tied in a characteristic way to associated "differential equations." Of course, since we are now dealing with *sets* of chronicles, we will generally expect to end up with *sets* of such "differential equations."

Let us sum up the essence of these strategies, involving three rather different ways of trying to cope with a nonrecursive chronicle f. As we shall see subsequently, they embody the three different approaches that have been taken in "system theory." Each of them requires something different of the auxiliary chronicles $g_1, \ldots, g_i, \ldots$. In the first strategy, we require only relative recursiveness, of f; i.e., a single diachronic entailment rule of the form

$$f(n+1) = T(f(n), g_1(n), \ldots). \qquad [4G.1]$$

In the second strategy, in which the g_i form a recursive set, and f is an observable, we require much more. In particular, since the g_i form a recursive set, we require by definition the diachronic entailment rules

$$g_i(n+1) = T_i(g_1(n), \ldots, g_i(n), \ldots) \qquad [4G.2]$$

for every argument n, and every chronicle g_i. Moreover, because f is an observable, we require by definition the *synchronic* entailment

$$f(n) = \Phi(g_1(n), \ldots, g_i(n), \ldots). \qquad [4G.3]$$

This is enough to entail the diachronic recursion rule for f itself:

$$\begin{aligned} f(n+1) &= \Phi(g(n+1)) \\ &= \Phi(Tg(n)), \end{aligned} \qquad [4G.4]$$

where we have abbreviated by $g(n)$ the set of values $(g_1(n), \ldots, g_i(n), \ldots)$ and by Tg the set of values $(T_1g(n), \ldots, T_ig(n), \ldots)$.

The final strategy, in which the chronicles g_i have the property that $\{f, g_1, \ldots, g_i, \ldots\}$ is a recursive set, gives us now the entailment rule for f,

$$f(n+1) = T_o(f(n), g_1(n), \ldots) \qquad [4G.5]$$

together with the rules

$$g_i(n+1) = T_i(f(n), g_1(n), \ldots), \qquad [4G.6]$$

once again valid for every argument n, and every chronicle g_i.

We may parenthetically remark here, for future reference, that the first strategy, embodied in [4G.1], treats the chronicles $\{g_i\}$ as externally supplied *inputs;* the second and third strategies regard them as representing a partial or complete set of *internal state variables.* And, of course, there is nothing to prevent us from combining these strategies.

I shall now show that these various generalizations of recursiveness, from individual chronicles to sets of chronicles, also allow us to directly generalize the notions of "differential equations," and of constraints, and hence, to formulate our new recursion rules in terms of them.

The trick in accomplishing this task, as is well known, is to reduce everything back to the case of a single chronicle, which we have already dealt with in detail. This, in turn, is accomplished by noticing that a set $G = \{g_1, g_2, \ldots, g_i, \ldots\}$ of chronicles, each member of which takes its values in a simple set, can be regarded as a single chronicle, taking its values in a more complicated set.

The formal operation necessary to accomplish this transmutation is, of course, the Cartesian product. For simplicity, suppose $G = \{_1, g_2\}$, where each element g_1, g_2 of G is a chronicle taking its values in numbers:

$$g_1, g_2 : \mathrm{Z} \rightarrow \mathrm{Z}.$$

Define a new kind of chronicle, which we shall call $g_1 \times g_2$, taking its values not in numbers but in *pairs* of numbers as follows:

$$(g_1 \times g_1)(n) = (g_1(n), g_2(n)).$$

Thus,

$$(g_1 \times g_2) : Z \times Z = Z^2$$

is a chronicle associating with each instant n a *pair* of values as indicated, and hence, a single point or element of the Cartesian product Z^2.

This Cartesian product operation can, of course, be generalized to arbitrary sets $g = \{g_1, \ldots, g_i, \ldots\}$ of chronicles. With an obvious notation, we define a new chronicle

$$\prod_i g_i : Z \rightarrow Z^{\mu G}$$

where μG is the cardinality of G, by writing

$$(\prod_i g_i)\,(n) = (g_1(n), \ldots, g_i(n), \ldots)$$

If now G is a *recursive* set, then there is a mapping

$$T : Z^{\mu G} \rightarrow Z^{\mu G}$$

such that

$$(\prod_i g_i)(n+1) = (\prod_i g_i)(n)$$

for every instant n, i.e., such that the diagram

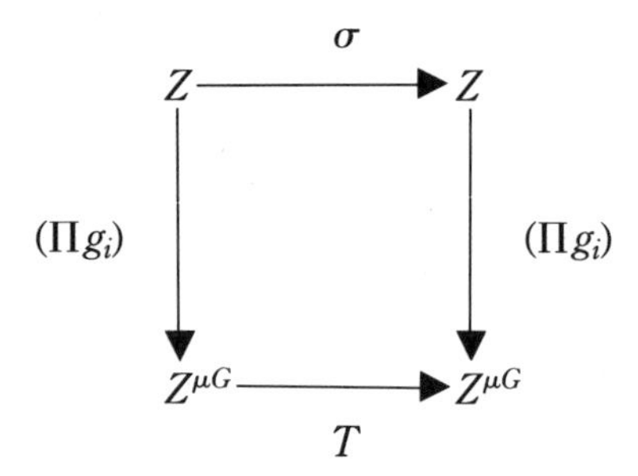

commutes.

But we have seen this kind of thing before; see figure 4D.1 above. Indeed, the entire discussion of recursion I have given before can be taken over essentially verbatim to the present situation. Thus, in particular, each initial choice of a point in $Z^{\mu G}$ generates a *trajectory* $\{T^n p_o\}$ *in* $Z^{\mu G}$, which is a curve; an image of the set of instants in $Z^{\mu G}$.

We thus have a lot of mappings

$$\varphi : Z \rightarrow Z^{\mu G},$$

of the form

$$\varphi(n) = T^n(p_o).$$

To each of them, Taylor's Theorem can be invoked, now in the appropriate multidimensional form. And we can therefore basically repeat the discussion we have already given in the one-chronicle case, to show that (1) the recursion rule T can be expressed in the form of a set of "differential equations," for which the individual chronicles g_i in G constitute solutions, and (2) there exist constraints, identical relations among the temporal derivations of the g_i; these constitute systems of *synchronic* "differential equations" that the individual g_i again satisfy.

Thus in all cases, the requirement of recursion does many wonderful things for us. Recursion, as we have seen, is a device for introducing entailment between values or entries in chronicles, more precisely, entailment between what is happening now and what will (in fact, what must) happen next.

So far, our discussion of recursion has been purely formal, belonging entirely to the inner world of formal systems or mathematics. To be sure, we started from the idea of a chronicle, but we started our analysis precisely by forgetting that the entries in a chronicle were encodings of observations of events in the ambience. Thus, a chronicle became entirely a formal object, a mapping from numbers to numbers. It is now time to return to our original situation, in which chronicles constitute *data,* encoded into a formalism from the ambience. When we do so, we will see that *recursion itself encodes something about the ambience.* What it encodes is not data, but a *property* of data; the assumption we make about *what* recursion encodes, and how that encoding is accomplished, in fact constitutes a new facet of Natural Law itself.

I shall therefore now turn to an analysis of Newtonian particle mechanics, from the point of view I have been in the process of developing. As I have asserted several times already, every mode of analysis of natural systems practiced in contemporary science finds its roots here. In particular, its modes of encoding and decoding have become so ingrained in us as to be invisible, in fact, have become synonymous with science itself. Yet as we shall see, Newton was not being entirely accurate when he said *"hypothesis non fingo."* We will find not one but many gratuitous hypotheses, which as I shall argue, continue to profoundly limit the scope of contemporary physics and serve in particular to make biology unreachable.

As I shall now show, Newton's laws are at heart nothing but Taylor's theorem, raised now to a universal principle via tacit but extremely restrictive limitations on encoding and decoding of events in the ambience.

4H. Newton's Laws

I have already alluded, several times, to the publication of Newton's *Principia* as the watershed event of modern science. Its influence has radiated out in many directions over the years, and I have talked about some of them in previous sections. Now I shall argue that the central concept of

Newtonian mechanics, from which all others flow as corollaries or collaterals, is the concept of state, and with it, the effective introduction of recursion as the basic underpinning of science itself.

Natural Law, as I developed it, does not in itself say anything about recursion, or about states, or even about systems. Newton does talk about all of these things. Indeed, he does so in such a way that Natural Law becomes an embodiment of mechanics, and hence of recursion. Thus, in my view, the *Principia* ultimately mandated thereby the most profound changes in the concept of Natural Law itself; in some ways a sharpening, but in deeper ways, by imposing the most severe restrictions and limitations upon it.

It is precisely with these restrictions and limitations that we shall be concerned. As we shall see, in the next few sections, they constitute in themselves not an inherent part of Natural Law, but rather embody only one special way in which Natural Law may be manifested. Nevertheless, when they are addressed at all, they are seen rather as an inherent, integral part of science itself. They have indeed *become* science.

Let us begin with some general remarks about Newtonian particle mechanics before we turn to the main part of our discussion. The first observation is that, philosophically, the *Principia* belongs to the tradition of Greek atomism, going back at least to the fourth century B.C. The atomists were *analysts,* seeking to break reality into some kind of ultimate units; first into "elements," and then analyzing these "elements." Atomism itself can be seen as articulating one of two mutually exclusive and all-encompassing alternatives; either the "elements" were infinitely divisible, or they were not. In the latter case, there was an inherent limit to analysis, and these were the *atoms*. Everything we see in the world of phenomena can thus be analyzed only so far, and no further.

The structureless particles of the *Principia* are thus essentially the atoms posited by the Greek analysts. But the *Principia* did not concern itself with analysis; it was instead a book about *synthesis*. It tacitly accepted as *given* the fruits of analytic atomism and addressed rather the question: what behavior can be manifested by such particles, individually or collectively?

If indeed analytic atomism is correct in assuring that *in the last analysis* there is "nothing but atoms and the void," then all of reality is necessarily embedded in the Newtonian synthetic theory. In this way, the synthetic theory *becomes* Natural Law, and that, in a nutshell, is precisely where we are, even today. The Newtonian synthetic theory has of course been extensively generalized, modified, and extended, as we have seen, but only in essentially technical directions; the heart of it has not changed. And the heart of it, as I will now argue, is *recursion*.

The Newtonian "atom," the particle or mass point, is characterized by the absence of anything "inside" it. It is *structureless,* its attributes incapable of further analysis. In fact, it *has* no attributes, other than its positions at successive instants, and whatever can be derived from these positions and their succession. (There are, of course, other numbers or parameters inherently to be associated with such a particle, e.g., its mass, but we will come to these later.)

It cannot be stressed too strongly that the Newtonian particle is a *formal object.* Tacitly, of course, it is *intended* as an encoding of something in the external world, or better, of something that *could* be in the external world, a "real particle." This is a crucial point, because the corresponding *decoding* is not at all obvious. We shall see what this means as we proceed.

The simplest (natural) system, for Newton, is thus a single "real" free particle, something that encodes into a structureless mass point. Aside from the parameters with which it is endowed (e.g., its mass), which we will take up in due course, its *only* attribute at any instant is its position in space. Thus, there is only one question to ask about it: namely, where is it? The answers constitute a chronicle, consisting of positions in space, labeled by corresponding instants of time. As with any chronicle, we can seek to extrapolate it back into the past and ahead into the future: where has the particle been? Where is it going to be? To answer these questions means to exhibit spatial position as an explicit function of time.

But as we have seen, merely exhibiting the values of a chronicle (in this case, positions of a particle in terms of the instants that label them) *involves no entailment.* It was part of the incredible insight underlying Newton's approach to recognize that what was necessary was more entailment, to realize that where a particle is going to be must be somehow entailed from where it is now. In a word, what is required to answer the question is recursion.

Newton's own researches into the differential calculus were undertaken, at least in large part, with these ideas in mind. He knew, for instance, that associated with any temporally varying quantity, such as the position of a moving particle, there was associated another quantity, which he called *velocity.* He identified this with the *temporal derivative,* the rate of change of position. At any instant, then, we now have two chronicles, the first consisting of particle positions, the second consisting of particle velocities, each labeled by the same set of instants.

It took amazing insight to realize that there existed *no entailment between the values* entered in the position chronicle *and the values* entered in the velocity chronicle. At first sight, this is very odd. It is odd because, in purely formal terms, any (differentiable) function *entails its derivative as another function.* But this entailment involves a complicated limiting process; it requires, not only a knowledge of the function's value *at* an instant, but also of *its value at every other nearby instant, past and future.* The upshot is that, although indeed a (differentiable) function entails its own derivative *as a function,* its *value* at any instant entails nothing about the value of the derivative at that instant.

Newton proceeded, in effect, to decode the derivative of the position of his formal, structureless particle into a new quantity associated with a "real" particle. Namely, he identified the formal temporal derivative of position with the "real" attribute of velocity, which can of course be measured. Thus, in the "real" external world, there were two chronicles to be now associated with a particle: its positions and the corresponding velocities. Because the *values* entered in the two chronicles are independent (i.e., no entailment exists between them), we seem to have made matters worse rather than better. For we now have *two* questions we can ask about our particle: "where is it?" and "how fast is it going?"

And it gets worse still, because the process leading formally from position to velocity can be iterated indefinitely. The next step is, of course, to differentiate the velocity. We can do this formally. But once again, Newton proceeded to *decode* this formal derivative by identifying it with another attribute of a "real" particle, its *acceleration.* Once again, acceleration represents a new chronicle to be associated with a "real" particle. And just as before, the individual values in this "acceleration" chronicle are independent of, and unentailed by, the corresponding values in the velocity chronicle and the position chronicle.

As I said above, this process can be iterated. Formally, we can differentiate the acceleration function, and then differentiate that, and so on *ad infinitum.* We thus formally associate with any formal moving particle an infinite number of temporal functions, and therefore, a correspondingly infinite number of chronicles become associated (decoded) as attributes of a "real" particle. And these chronicles are all independent; there is still no entailment between their entries.

Nevertheless, there is indeed an entailment now available to us from this infinity of chronicles, which we have seen before, namely

$$x(t_o+h)=x(t_o)+hx'(t_o)+h^2/2!x''(t_o)+\ .\ .\ .$$

where $x(t)$ represents the position of our particle at the instant t.

In fact, it will be noted that the set of chronicles $\{x^{(k)}(t)\}$, forms a recursive set. In addition, this set constitutes all there is to know about the formal Newtonian structureless particle, and hence, all that is knowable about the "real" particle it encodes.

We will say that the set $\{x^{(k)}\}$ constitutes a set of *state variables* for our formal particle. The specific set of values $\{x^{(k)}(t)\}$, which these functions assume at the instant t will be called the *state* of the particle at that instant. Being a recursive set, all we need to do is to know, or find out, or measure the values of each of these state variables at a common instant of time (say t_o), and Taylor's Theorem does the rest.

As it stands, though, the state of a formal Newtonian particle is an infinite object. It represents an eerie anticipation of ideas that would become current only some three centuries later, with the advent of quantum mechanics (to which it is in fact closely related; see section 4J below). But such ideas were not congenial in 1680, nor even in 1880, and so something more was needed to turn the state concept, and the recursion it embodied, into something more usable.

So far, we have been looking only at the single particle that constitutes our (natural) system. Let us now notice that the same act that singled out that particle from the ambience also *simultaneously* specifies the system environment. Newton thought of the actual motion, the change of position with time, of a "real" particle as arising from its interaction with (or better, its reaction to) whatever constituted the environment. But if the state of an individual particle is at this point already an infinite object, to try to characterize a "state" of this environment is literally unthinkable. Newton did not even attempt this. Instead, he took a completely different approach, embodied in the concept of *force.* The environment was to be characterized, not in any kind of absolute terms, but rather entirely through its effects on the *system.* Specifically, we characterize the environment of a particle through its effect on change of state.

Newton chose an ingenious device for doing this. In effect, he noted that *at the purely formal level,* and *only* at that level, we are free to make the environment of a Newtonian particle anything we want. We can even make it consist of nothing at all. This is a situation without counterpart in the external world. But in formal terms, it establishes a reference, against which the effects of more general environments can be estimated.

The substance of Newton's First Law is to specify what a particle does in such an empty environment. He mandated that *such a particle cannot accelerate:*

$$x''(t) \equiv 0.$$

This law thus takes the form of a very simple *constraint.* It is, of course, also a differential equation, which must be satisfied by the position chronicle $\{x(t)\}$. Formally, all we need to know to specify this part are two numbers, which may be taken as initial values of the position chronicle and of the velocity chronicle of the particle. The constraint thus assures us that we do not need to know the full, infinite state of the particle; our infinity of chronicles has collapsed to only two.

In a formal environment that is not empty, and is thus the seat of *forces* impinging on the particle, the effects of these forces on the motion of the particle is thus tied to its *acceleration.* The way in which force and acceleration are explicitly tied together constitutes Newton's Second Law, which consists of two separate parts. In the first part, which is generally regarded as "the Second Law," Newton mandates that the effect of any environment forces on a particle, at any instant, are proportional to the particle's acceleration at that instant;

$$F = m\,x''.$$

Here, the constant of proportionally m is identical with the mass of the particle, one of those parameters or attributes inherent in the particle, which we mentioned before. Intuitively, this parameter m represents how the particle couples with, or responds to, what is going on in its environment.

The second part of Newton's Second Law says essentially the following. When the force F, which pertains to the particle's environment, is thus reflected in the particle's own behavior, then that force depends *only* on where the particle is, and on how fast it is going, at any instant of time. In other words: from the particle's point of view, any forces F it experiences from its environments *are already functions of its own states,* in fact, functions of position and velocity *alone.*

Thus, putting all this together, Newton's Second Law asserts the following: any environment with which a formal particle interacts is described by a function F (the force it exerts on the particle) such that, at every instant t, we have

$$F(x(t),\ x'(t), t) = m\,x''(t). \qquad [4H.1]$$

Let us pauses for a moment to mention the following technical point. If the function F, which describes the environment, does not in fact depend explicitly on the time t, the particle is called *autonomous;* otherwise it is *nonautonomous.* Autonomy is thus a condition on the environment; if it holds, then Newton's Second Law for autonomous particles says

$$F(x(t),\ x'(t)) = m\,x''(t) \qquad [4H.2]$$

for every instant t. But this is just a *constraint,* i.e., an identical relation satisfied by the chronicle $x(t)$ and its temporal derivatives. It is also a differential equation of the second order, which must be satisfied by the chronicle $x(t)$ telling us where our particle is at every instant of time.

Since the time of Newton, physics has been almost exclusively concerned with the autonomous case, the one for which Newton's Second Law takes the form [4H.2]. The constraint it expresses constitutes the *equation of motion,* or *dynamical equation,* of the particle. This is in fact the cutting edge of the recursion, the point of it all.

Since these ideas have been so important, I will pause here to review and sum up what I have said so far.

I began by partitioning the ambience, the external world, into a *system* (in this case, something encoding into a single formal particle) and its *environment.* By definition, the only temporally varying attributes of the system are the position of the particle at an instant (a chronicle $x(t)$), and the temporal derivatives of that position (the chronicles $x'(t)$, $x''(t)$, . . .). the values of these chronicles at any instant of time collectively express everything there is to know about the particle at that instant; they determine its *state.* However, Taylor's Theorem tells us that the state is *recursive;* present state *entails* subsequent state. This is the essence of the concept of state, its recursive character.

The upshot of Newton's Second Law is to effectively collapse the state of the particle, which is an infinite set of variables, *down to only two* of them, down to position and velocity alone. This collapse proceeds in several steps. First is the representation of the environment entirely by its effect on the system, in this case, by the force it exerts on our particle. Second is the reflection of this force, and thus the environment it represents, in a single chronicle pertaining to the particle alone, namely, the acceleration $x''(t)$. And finally, in the autonomous case, the expression of this force as an explicit function of $x(t)$, $x'(t)$ alone, leading to the constraint expressed in [4H.2], the equation of motion.

Following long custom, I shall call the pair of chronicles $(x(t), x'(t))$ the *phase* of the particle. [4H.2] says that *this is already a recursive set of chronicles;* from these, all the other chronicles $x^{(k)}(t)$ can be effectively entailed, via the constraint [4H.1].

Newton's Laws thus serve to transmute the initial dualism between system and environment into a new dualism, that between *phase (states) and forces,* or between *states and dynamical laws.* The states or phases constitute a description of system. Environment, on the other hand, gets an entirely different kind of description; it is described in terms of the specific recursion rule it imposes on states or phases. It is this dualism between states and dynamical laws that, more than anything else, has determined the character of contemporary science. The encodings (and even more, the decodings) embodied in it have indeed, as I have pointed out several times, become essentially synonymous with science itself.

I must say a word about the generalization from a single formal Newtonian particle to many-particle systems, because an essential new idea is involved. In the one-particle case, there is no prescription allowing a particle to itself *exert* forces; in our picture so far, a single particle can be pushed around only by forces impressed on it by its environment. This purely reactive character of a single particle is very deeply embedded in the concept of state itself; in the Newtonian picture, a particle is shoved from one state to another through the external agencies to which it must react.

But when we go to systems of several formal particles, the situation is different. Each particle forms part of the environments of all the others; thus, in addition to getting pushed around by the forces impressed upon it, each of these particles must be capable of doing some pushing on its own. This, as I have said, requires a new ingredient in the Newtonian picture, an ingredient that allows a particle to be a seat of force, as well as the target of force.

Hence, the separate stipulation of, for example, the Law of Universal Gravitation: any particle *exerts* a force on any other particle, directly proportional to the produce of their *masses* (parameters, state-independent) and inversely proportional to the square of the interparticle *distance* (state-dependent).

Thus, in assessing the motion of particles belonging to multiparticle systems, we can now distinguish between two kinds of forces: (1) those arising from the other particles in the system (the *internal* forces) and (2) those impressed from the environment of the system as a whole (the *external* forces). The status of these two kinds of forces is, of course, quite different. Specifically, the internal forces now receive *two* quite different kinds of description. On the one hand, through the stipulation of how the constituent particles of the system exert forces, we can represent the internal forces as definite functions of the states of the system as a whole. On the other hand, the collective internal force that each particle in the system sees is reflected as a (generally nonautonomous) function of its own state alone.

Thus, the internal forces possess a peculiar reflexive character, on which are superimposed the effects of forces impressed from the environment. If such a multiparticle system were in an *empty* environment, so that all the forces were internal forces, there is a sense in which the system would *push itself* through its own states. This is the *free behavior* of the system. Just as in the one-particle case, it has become customary to employ this free behavior as a standard, against which the effects of impression external forces from a nonempty environment may be compared.

With these provisos, the formal treatment of multiparticle systems follows that of single-particle systems. Each particle in the system is endowed with position and the temporal derivatives of position. The totality of all of these, by Taylor's Theorem, constitutes a recursive set. Just as in the one-particle case, we impose a set of constraints of the form [4H.2], which now constitute the equations of motion, or dynamic laws, imposed on the phases of the system as a whole. The details of doing this constitute the substance of every book on classical particle mechanics.

Thus, just as in the single-particle case, the system-environment dualism becomes transmuted into a dualism between phases and forces. It is a little more complicated now, because at least some of the forces are internal. But insofar as the environment affects the behavior of the system, its only description arises through the specific form of the equations of motion superimposed on the system phases, or states, as we have seen.

Two more remarks and we are done with this brief review of the Newtonian picture.

The first remark concerns nonautonomous systems, systems of particles for which the forces imposed by the environment on the particles of a system contain time explicitly. In this case, the recursion rules governing change of states or phases of the system become *relative.* We have seen this situation before (see section 4G above); additional chronicles $\{g_i(t)\}$ must be stipulated, which pertain to the *environment.* Once these new chronicles are stipulated, we recover a set of recursion rules in the form of nonautonomous differential equations containing the $g_i(t)$. In the parlance of control theory, these $g_i(t)$ are generally referred to as "forcings," or "inputs," and the emphasis is shifted from change of phase as a function of *time* to change of phase as a function of *inputs.* But at root it is exactly the same picture.

The second remark pertains to my use of the term *constraint* to express any identical relationship satisfied by a chronicle $x(t)$ and its temporal derivatives. Historically, the term arose in connection with the conversion of Newtonian particle mechanics into a form more suitable for dealing with bulk matter, matter that, from an atomic point of view, must consist of enormous numbers of individual particles. In formal terms, this conversion was effected by reducing the dimensionality of phase spaces. The only way to do this is to stipulate identical relations that must be satisfied by the otherwise independent coordinates of these big phase spaces. These coordinates, as we have seen, can be regarded as the positions of constituent particles and their associated velocities. As Helmholtz showed, some two centuries after the *Principia,* such identical relationships could be expressed locally as *linear* relationships identically satisfied by the coordinates of position at an instant and their differentials (essentially the corresponding velocities), as the vanishing of a differential form. These kinds of identical relationships were called constraints; rigidity, for example, is a familiar kind of constraint in this sense. I have simply generalized the term; instead of connoting merely an identical relation between chronicles and their first derivatives (e.g., between coordinates of phase), I employ the term to denote *any* identical relation between a chronicle and *any* of its temporal derivatives. Thus, for us, Newton's Second Law is itself a constraint. As we have seen, constraints are always associated with recursion relations (in the form of differential equations) and my use of the term makes this association clear in general.

41. On Entailment in Physics: Cause and Effect

Any system, be it formal or natural, is characterized by the entailments within it. Thus, a *formal* system is characterized by its inferential structure, which entails new propositions from given ones. And a *natural* system is likewise characterized by its causal structure. Natural law serves precisely to connect the two kinds of entailments, by asserting that they can be brought into congruence (via suitable modes of encoding and decoding). In the present section, we shall examine in detail the entailment structure of Newtonian particle mechanics.

It must be stressed at the outset that Newtonian mechanics is a *formalism*. It is thus characterized entirely by an inferential structure, and as such, needs, of course, to have no external referents at all. Formalisms become science, as we have seen, precisely when their elements are endowed with referents in the external world, more specifically, when they are inserted on the right-hand side of a modeling diagram like figure 3H.2 above. Indeed, they acquire their external referents by virtue of the specific encoding and decoding arrows that have been mandated. When a modeling relation exists between the inferential structure of the formalism and that which, in the ambience or external world, has been coded into it, that modeling relation expresses a congruence between inferential entailment in the formalism and causal entailment in what has been encoded. Thus if we believe, as we have for centuries past, that the Newtonian formalism actually *models* the external world, the characteristic inferential structure in Newtonian systems says something about causality, indeed, about causal entailment in general.

As I shall now show, Newtonian mechanics as a formalism manifests a surprisingly weak inferential structure, in the sense that *almost everything of importance in it is unentailed.* When we translate this back to the language of *causal* entailment, we will find *the province of causality correspondingly restricted.* This poverty in entailment, which is intrinsic to the formalism itself, seems at first sight to make that formalism very general, in that many diverse things can be encoded into it, *but the paucity of entailment makes the formalism very special as a formalism.* And if a formalism is special, *what makes it special is automatically impressed on anything encoded into it;* it will thereby come to look special too. And that is the crux of it.

I must point out that the decisive impact of Newtonian mechanics on scientific thought, for good and ill, was not Newton's fault, any more than, say, the proliferation of divergent series in mathematics prior to Cauchy was Euler's fault. Newton himself was concerned only with what happens in formal systems of structureless mass points, which he himself applied mainly to problems of celestial mechanics. His success in these endeavors was so great that an infinite credibility was conveyed upon his methods per se. Historically, this success occurred at a most propitious time; the latter part of the seventeenth century was a time of boundless optimism and prosperity in much of Europe, with a concomitant pride in the capacities of human action and human thought. Newton vindicated and nourished the faith in unlimited human progress characteristic of those times. It is no surprise that his methods were thus elevated to universal proportions by his contemporaries and their successors. Moreover, his ideas had the virtue of giving everyone plenty to do, a universal framework for the theorist to play with, and at the same time, infinite room for the experimentalist, the observer, and the analyst. For if understanding reality was now a matter of recipe, of finding the ultimate atomic particles and determining the forces acting on them, these were now *empirical* problems mandating empirical solutions in the form of data. Only biology, to its chagrin, was forced to stand aside from this triumphant wave, at least until our own century; indeed, molecular biology can be thought of as a retarded impact of that wave, which washed over the rest of the intellectual world three centuries ago.

I turn now to a discussion of entailment in Newtonian systems. There are two aspects to this, which must be carefully distinguished: (1) entailment in the Newtonian *formalism,* apart from any external referent, which is necessarily inferential; and (2) causal entailment, which is imputed to the ambience once the formalism is endowed with external referents, by virtue of specifying encoding and decoding rules. I will consider the formalism first, then the (canonical) Newtonian encoding and decoding procedures, and finally, what these assert about causal entailment in the ambience.

On the purely formal side, I have already explained how Newton's

Second Law serves precisely to entail subsequent phases from present phases. In general multiparticle systems, the role of position is now played by *configuration.* The configuration of such a system is specified by giving the positions of its constituent particles. *Velocity* now means temporal derivative of configuration and is correspondingly specified by giving the velocities of the constituent particles. A pair consisting of a configuration and a corresponding velocity is a *phase.* Just as in the single-particle case, Newton's Second Law ensures that phase is recursive.

Let us write out the formal recursion rules, which govern the succession of phases. Suppose we are initially in a particular phase

$$(x(t_o),\ v(t_o)).$$

For each h, we want to specify the operation

$$T_h\colon (x(t_o),\ v(t_o)) \rightarrow (x(t_o+h),\ v(t_o+h)).$$

Newton's Second Law in this case asserts that there is a function $F=F(x,v)$ *of phase alone* such that

$$(x(t_o+h),\ v(t_o+h)) = (x(t_o)+hv(t_o),\ v(t_o)+hF).$$

From this expression, which is nothing but Taylor's Theorem, it is clear that in general the family $T=\{T_h\}$ forms a one-parameter group; each T_h is a $1-1$ mapping of the space of phases onto itself, and we always have

$$T_hT_g = T_g + h.$$

Thus, the image

$$T(x,v) = \{T_h(x,v)\}$$

of any phase (x,v) constitutes a *trajectory* of phases, governed by the recursion rules T_h. These trajectories obviously also satisfy the constraint that constitutes the equations of motion of the system and that is expressed in Newton's Second Law. In fact, all we have done is to formally rewrite that constraint as an explicit set of recursion or entailment relations.

In formal terms, then, these recursion rules T_h express the full inferential structure of the Newtonian formalism. There is no other entailment present in that formalism, nothing but the entailment of subsequent phases from present phases.

This entire picture, I stress again, is at this point pure formalism, independent of any external referents. The use of the word "particle" for the basic element under discussion is so far purely gratuitous. The formalism will not become science, will not become physics, until specifically endowed with such external referents, by virtue of specific encoding and decoding arrows, and the establishment of a corresponding diagram of the form [3H.2].

Of course, historically, Newton was actually building his formalism with specific referents in mind. He thus did not separate the encoding/decoding aspects from the purely formal ones. We, on the other hand, must separate them, because we are interested in the imaging of causality by means of a formal inferential structure; this cannot be reached by looking only at the formalism qua formalism.

Newton thus began from the tacit presupposition that there were things in the ambience that could be encoded into his structureless formal particles. He was doubtless thinking not so much of *atoms* in the Greek sense as he was of celestial objects, whose spatial extension was negligible compared to the distances between them. In any case, he proceeded to identify the formal attributes of his particles with certain *chronicles* pertaining to such external objects. Since the only formal attributes available are position and its temporal derivatives, we only need to assign or encode particular chronicles into these. And since the Second Law formally tells us that all these temporal derivatives are determined if *phase* is known, we actually only need to encode into position (configuration) and its first temporal derivative.

Well, position is straightforward; if there is a "real" particle in the ambience, its positions can be chronicled, relative to some suitable reference frame (absolute space). Hence configurations of families of particles can also be chronicled. These of course encode into paths in configuration space, parameterized by time (absolute time). Temporal derivative of configuration is more difficult. Newton supposed offhand that it encodes *velocity*. But there are problems with this encoding of measured and chronicled velocities into temporal derivatives of configuration, some of which lie at the heart of the transition to quantum mechanics.

We then come to temporal derivative of velocity, which encodes *acceleration* in the Newtonian scheme. When thus encoded, acceleration has three distinct roles to play. First, it is an independently, measurable quantity, whose values can be chronicled quite independent of anything else. The Second Law says, however, that the *values* of acceleration at any instant are determined by the values of phase at that instant; this is the constraint or identical relation that expresses the equations of motion. And finally, the Second Law says that the acceleration, or more precisely, the *way* in which acceleration depends on phase, is a measure of the *force* impressed on the particles of a system *by its environment.*

Thus, acceleration is a busy quantity. On the one hand, it simply encodes a chronicle pertaining to *system.* But the Newtonian formalism requires this acceleration chronicle to be synchronically related to phase, an observable of phase. This generates the recursion rules that entail subsequent phases from present phases, as we have seen. But the *way* in which acceleration values are determined from phase values, or in formal terms, the specific *form* of the function $F(x, v)$, also now encodes something. *What* it encodes is a question that Newton boasted he never had to ask, but whatever it is, it is something about *environment.*

Thus we see explicitly what I mentioned before, the dualism between system and environment transmuted into a new dualism between *states* and *dynamical laws.* States pertain to *system,* and the succession of states that is characteristic of system constitutes chronicles. Dynamical laws take the form of *identical relations* between certain kinds of chronicles; they generate the recursion rules that govern the diachronic entailment of subsequent state by present state, but *the relations themselves* constitute an encoding of environment and not system. Thus, in the Newtonian picture, systems get states; environments do not (and cannot); environments rather become identified with dynamical laws, i.e., with the rules governing the diachronic succession of states.

This is a fateful situation. Once we have partitioned the ambience into a *system* and its environment, and (following Newton) once we have encoded *system* into a formalism whose only entailment is a recursion rule governing state succession, we have said something profound about causality, and indeed about Natural Law itself. In brief, we have automatically placed beyond the province of causality anything that does not encode directly into a state-transition sequence. Such things have become *acausal,* out of the reach of entailment in the formalism, and hence *in principle undecodable* from the formalism.

Let us see what all this does to Natural Law. As I formulated it in section 3H above, it asserted only that (1) there were entailments (causal relations) between events in the ambience, and (2) these could be put into congruence with inferential entailments in formalisms. In this usage, "event" is a necessarily vague word, the kind of word that cannot be defined but only interpreted. Crudely, and necessarily informally, we can synonymize the term "event" with anything we can ask "why?" about.

In the Newtonian language, "event" is given a much sharper interpretation. It is now synonymous with "system N is in state x. *Why* is system N in state x? Well, because (1) system N was earlier in some *initial* state x_o; (2) because the system parameters, which mediate between N and its environment, have certain values; and (3) because the environment itself imposes a recursion rule on the states of N, determining a trajectory through x_o, which passes through x.

This constitutes, of course, a paraphrase of the decoding of a formal Newtonian inferential structure into a causal language. We have seen it before (see section 3G above), as I indicated, we find here the ghosts of three of the four old Aristotelian categories of causation, namely, (1) material cause, (2) formal cause, and (3) efficient cause, respectively. But now, Aristotelian causality has collapsed down to state entailment in systems, because that is all there is in the formal Newtonian language to be decoded back into causal terms at all.

Actually, the situation is even more circumscribed than this. If, as is common, we suppose that every system N is in fact a collection of structureless particles, then "event" becomes synonymous with *phase* of that collection, and the recursion rule governing change of phase becomes Newton's Second Law. From this it follows directly, as I have said before, that every natural system N possesses a *largest model* (namely, the one I have just described), from which *every other model* can be effectively extracted *by purely formal means.* Reductionism can be characterized as the search for this largest model. Furthermore, I repeat that this largest model is of an essentially syntactic nature, in that structureless, unanalyzable elements (the particles) are pushed around by mandated rules of entailment that are themselves beyond the reach of entailment.

For future reference, I shall define a natural system N to be a *mechanism* if it possesses the properties I have just articulated: namely, (1) it has a largest model, consisting of a set of states, and a recursion rule entailing subsequent state from present state; and (2) every other model of it can be obtained from the largest one by formal means. The reformulation of Natural Law that I have just given then boils down to the assertion that *every natural system is a mechanism.* I shall return to this assertion later, when I discuss *machines,* especially in chapter 8 et seq.

4J. Quantum Mechanics, Open Systems, and Related Matters

As I have already indicated, several basic presuppositions of Newtonian mechanics break down when extrapolated from experience with bulk matter to the realm of the small. That is, when we come to actually think about *atoms,* the Newtonian encodings fail completely. This has had several revolutionary consequences in physics. One of them is the development of a new kind of mechanics, *quantum mechanics,* which I have already mentioned several times (see section 2E above). Concomitant with this has been a proverbial agonizing reappraisal of causality itself, which is worth mentioning here. As we shall see, as far as causal entailment is concerned, the quantum-theoretic revolutions were mainly technical; the heart of Newtonian causality (recursion) has passed intact from classical to quantum mechanics.

Until the advent of quantum ideas, physicists did not in fact think much about causality; as far as they were concerned, the question had long been settled along the lines I have described above. Indeed, Bertrand Russell (an otherwise perspicuous commentator) wrote a provocative essay, suggesting the concept of causality was superfluous and should be expunged from the scientific lexicon:

> . . . the word "cause" is so inextricably bound up with misleading associations as to make its complete extrusion . . . desirable . . . oddly enough, in advanced sciences such as gravitational astronomy, the word "cause" never occurs. . . . The reason why physics has ceased to look for causes is that, in fact, there are no such things. The law of causality . . is a relic of a bygone age, surviving, like the monarchy, only because it is erroneously supposed to do no harm.

Yet when quantum mechanics seemed to contradict or preclude classical ideas of causality, an enormous disquiet was generated, which has still not been completely resolved.

The problem is that the Uncertainty Principle, or more generally, the commutation relations on which Heisenberg based his quantum theory, are not compatible with the notion of *phase*. As we have seen, phase is the basic idea in the Newtonian description of particulate systems; it is precisely what the recursion rules operate on to generate the trajectories that encode causality in that formalism. The Heisenberg commutation relations said that classical phase could no longer even be defined at the quantum level, let alone be recursive.

But, as was quickly realized, giving up the notion of phase did not mean giving up the notion of state. It merely required an encoding of that notion into a more complicated mathematical or formal object *(wave function)* whose relation to actual observational chronicles was now (to say the least) indirect. Formally, in quantum mechanics, the wave functions that encode state remain completely recursive, governed now by Schrödinger's equation (or its equivalents) rather than by Newton's Second Law. The guts of classical causality therefore passed intact to the new mechanics. It so happened that the new *encoding,* into a formalism of wave functions and Schrödinger's equation, could be related only in a statistical way to the old, classical *encoding,* so that the two *inferential structures in the formalisms* could not be brought into a complete homology. But as we have seen, this is an entirely different matter; causality encodes differently into the two kinds of formalisms, but that only says something about the *encodings,* and not about causal entailment itself; see section 8F below.

Nevertheless, the reappraisal of causality occasioned by the advent of quantum theory has left physicists without consensus on what causality is or on how it should be encoded into contemporary physical formalisms. More generally, no one is today sure *what* the formalism of quantum theory encodes, or even if it encodes anything at all; in this latter view, advocated by Bohr under the rubric of complementarity, the only thing that matters is the *decoding.* I believe it fair to say that the "foundations" of quantum theory remain a quagmire, to a far greater extent than has ever been true in physics before.

It would therefore be idle, as well as perhaps presumptuous, to enter into a more detailed discussion of quantum theory here. My main point is, however, unarguable: that the concept of *state* plays the central role in its formalism, just as it did in its classical predecessor, and the essential property of state is its recursiveness. It thus perpetuates the duality between states and dynamical laws that began with Newton. The inferential or entailment structures in the two formalisms are different enough so that they cannot be directly compared (and indeed, attempts to directly compare the two *formalisms* have created much of the confusion to which I alluded above), but they remain different species of the same genus.

Let us now turn briefly to another matter connected with these formalisms. As I argued above, they serve to replace the vague word "event" with the apparently more precise "state x of system N", and ultimately, to replace "system N" by "family of structureless particles." This last replacement, which as we have seen is at the heart of reductionism, basically constitutes a redefinition of the term "system" (more specifically, of "natural system"). Indeed, it says that the terms "natural system" and *mechanism* are to be synonymous. The consequences of this constitute chapters 8 to 10.

Yet even if we accept the heuristic Newtonian view of the ambience as a sea of moving structureless particles, there are plenty of situations that we familiarly address in terms of systems but that are not in accord with this identification. For instance, suppose we are interested only in a delimited region of physical space, and hence, only in the particles that happen to inhabit this region. The only question here is "how many particles are in the region now?" and the answer is a number (say the number of particles per unit volume in the region; i.e., a density). This is a very natural question to ask, but it is one that already falls outside the reductionistic Newtonian framework. It does so, because the particles in the region do not comprise a definite, fixed family. In general, particles are entering the region, and then leaving it again, being spewed up from sources and disappearing down sinks. We do not therefore follow the same particles forever, but only during their residence times in our distinguished region of space. We thus no longer have *either* a fixed family of particles or a fixed environment.

A situation of this kind, where there is turnover, is called *open*. If we allow this kind of thing to be a *system*, there is clearly going to be trouble, yet there will also be trouble if such situations are excluded.

Over the years, physics owes much of its respectability to its success in resolutely avoiding direct confrontations with such open situations. Where confrontation could not be avoided (e.g., in fluid dynamics), it was noted that a Newtonian *language*, with a partition into "states" and dynamical laws, could sometimes encode the situation, if we did not look too closely at it. Specifically, the "state" was now to be a density of some kind, and the "velocity," or rate of change of state, was now specified, not in terms of force, but solely by bookkeeping. Namely, the change of "state" (density) during some small time interval is given by

(initial density)
+ (particles entering region during interval)
− (particles leaving region during interval).

But the last two terms pertain to the *environment*. In general, they should be represented formally as arbitrary functions of time. But if we make strong enough *hypotheses* about the environment, we can accomplish what Newton's Second Law did for particle mechanics: namely, reexpress these environmental terms as functions of the *internal* density (plus enough new parameters to give the appearance of a closed autonomous system of differential equations, and hence, to make this internal density again *recursive*).

Conceptually, then, the "open system" was regarded as sandwiched between two conventional Newtonian mechanisms. The lower part of the sandwich was composed of our fixed region of space, with some particles in it, but in an empty environment, e.g., surrounded by a wall isolating it from the rest of the world. The particles isolated by this impermeable wall constitute, of course, a straightforward Newtonian mechanism. Another, larger mechanism was obtained by, in effect, moving the walls farther out, to get a bigger, but again fixed, family of particles. But the "opening" of the smaller system in this way had to be done very, very cautiously, or else the notion of *state* of the system thereby opened would be lost; nothing recursive would remain.

Thus it is that physics has always had serious difficulties with "open systems," e.g., families of particles that can turn over. As we have seen, the difficulty is precisely in assigning to such a situation a notion of state that is recursive. The physicist has always tried to avoid dealing with such situations; where it could not be avoided, the strategy has generally been: (1) to close the system up again, obtaining thereby a set of states governed by recursion rules, and then, either (2) impose enough hypotheses to make the "open system" governed by new recursion rules on that same set of states, or (3) cede autonomy altogether and superpose on the recursion rules governing the closed system some explicit but arbitrary functions of time.

In every case, the strategy is then to regard the "open system" as an underlying closed system *plus something.* It is from this *closed* system that we extract the states and recursion rules that allow us to use the word *system,* consistent with Newtonian parlance. With these tacit provisos, the "open system" becomes *open system,* admissible without blatant contradiction, as system, into the Newtonian fold.

I shall return to this situation later. I merely remark here that at the present moment, there is still no "physics" of open systems. Largely, this is because of the insistence on thinking of an open system as only a closed system with some additional terms in its dynamical laws. Even if we accept this, it turns out that the dynamical laws that can encode the physicist's closed or isolated systems are very special, nongeneric. When such a system is "opened," its properties depend primarily on *how* it was opened, and not on the properties it possessed when it was closed.

Moreover, we should recognize the possibilities that the "opening" of a closed system cannot entirely be encoded as a change of dynamical laws (recursion relations), with the states left intact. This apparently innocent remark is actually of a very radical character and contains the seeds of much of our subsequent discussion; see especially section 9D et seq. below. In these connections, it will be helpful to keep the precedent of quantum theory in mind; it is precisely in reencoding the concept of state that the transition from classical to quantum mechanics manifests itself.

Chapter 5

Entailment Without States: Relational Biology

NOW, AFTER our extended detour into the world of natural systems and the formalisms that describe them, we are better equipped to move back toward biology. We shall make our approach to biology from an unfamiliar direction, one that most biologists do not recognize, in fact, a direction that seems to them to carry us away from what they understand as biology rather than toward it.

5A. A New Direction

We do this in part because it is actually inherent in what I have already said that there is no point at all in trying to approach biology from the familiar directions. That is, not if one truly wishes to cope with the fundamental question, "What is life?" Contemporary biology has concerned itself almost exclusively with the endlessly fascinating epiphenomena of life, but the secrets are not to be found there, no more than one can fathom the nature of the chemical bond by staring at the periodic table. Thus, we *must* approach the problem from a new direction; we take the particular one we do because, as I will argue, it is the right direction.

In section 2E above, I contrasted "contemporary physics" with the "ideal physics" it aspires to be. I argued there that a vast discrepancy exists between the two, a discrepancy most blatantly revealed by the inability of contemporary physics to throw light on organic phenomena. The comparable relation between "contemporary biology" and "ideal biology" is far more discrepant still. At the moment, biology remains a stubbornly empirical, experimental, observational science. The papers and books that define contemporary biology emanate mainly from laboratories of increasingly exquisite sophistication, authored by virtuosi in the manipulation of laboratory equipment, geared primarily to isolate, manipulate, and characterize minute quantities of matter. Thus contemporary biology simply *is* what these people do; it *is* precisely what they say it is.

On the other hand, a science indifferent to its own basic questions can hardly be said to be in its ideal situation, or indeed, anywhere near it. If the status quo is to be changed, we had better not entirely vest in the contemporary biologist the right to say what biology *is* in the abstract. For whatever biology will be tomorrow, it will not be merely an extrapolation of what it is today.

So our direction of approach to biology will be, in this sense, an unfamiliar one, which has been called *relational biology*. I will describe it in some detail, from the general standpoint of Natural Law; that is, I will be concerned with its formal aspects, and with what is encoded into and decoded from the formalisms.

On the formal side, we shall see that the inferential structure characteristic of relational biology is much richer than, and at the same time very different from, the formalisms we have considered heretofore. Our systems are assigned *no states, no environments, and there is no recursion.* Nevertheless, all of these are recaptured, as we shall see, by very special instances of the relational formalisms.

In later sections, we shall make direct contact between the relational formalisms and the phenomena of physics and biology, through the idea of *realization.* We shall see thereby that the causal entailments, in both biology and physics, reflected in the relational formalisms are much richer than those captured by states and recursion. We shall in passing acquire a much deeper insight into the machine metaphor and its various inadequacies. And then, of course, we shall investigate what these new directions tell us about the basic question "What is life?"

5B. Nicholas Rashevsky

The term *relational biology* was coined by Rashevsky in 1954, in connection with ideas that had been germinating in his mind for some time. I will digress at this point to give some of the history of those ideas and of the personality behind them. These are interesting and important in their own right and will be useful when I turn to articulate the ideas themselves. Moreover, this history merits a statement, to which Rashevsky is entitled.

Rashevsky described himself as "a stubborn Ukrainian." In fact, he considered the Ukrainians to be the most stubborn people in the world, and himself the most stubborn of these; this alone explains a great deal about him. He was born near Kiev in 1899 and thus grew to maturity in the most turbulent of times. Nevertheless, he took a doctorate in theoretical physics while barely out of his teens, and soon became a regular contributor to international physics journals like the *Zeitschrift für Physik;* he quickly published a score of articles dealing mainly with thermodynamics, relativity, and quantum theory.

In the late 1920s he found himself at the Research Laboratories of the Westinghouse Corporation in Pittsburgh, Pennsylvania. He was then working, among other things, on the thermodynamics of liquid droplets. Specifically, he found that such droplets became unstable past a certain critical size (i.e., when surface tension became too weak to offset diffusional and other forces impinging on the droplets) and would then spontaneously divide into two or more smaller droplets. In modern terminology, he was studying a bifurcation phenomenon.

As he related it to me, he met a biologist from the University of Pittsburgh at some social occasion. In the course of casual discussion, he asked the biologist whether the thermodynamic mechanisms on which he was working was the way biological cells divided. He was told (1) nobody knew how biological cells divided, and moreover, (2) nobody *could* know how biological cells divided, because this was biology. Rashevsky, who was always more of an "ideal" physicist than a "contemporary" one, found this outrageous, that a material phenomenon (the division of biological cells) should be so casually put outside the pale of physics.

This chance event turned Rashevsky to the study of biology. More particularly, it turned him toward establishing the material basis of basic biological phenomena in general. This was at a time when there was almost no precedent for attempting such a thing and no tradition to be part of; he was entirely on his own. But he was always clear in his own mind about what his aspiration was: in his own words, it was "the building-up of a systematic mathematical biology, similar in its structure and aims to mathematical physics."

Rashevsky turned completely to this task in the years that followed, armed with his formidable intellectual equipment and his unlimited reserves of energy and stamina. At first alone, and then with the aid of students and colleagues at the University of Chicago, he thanklessly blazed trails along which unknowing hordes plod today. Rashevsky concentrated initially on phenomena whose material basis seemed, if the pun will be forgiven, closest to the surface, cell division and excitability. As to the former, by the mid-1930s he had discovered and investigated the interplays between chemical reaction and physical diffusion that nowadays give employment to so many. Among other things, he discovered at that time the destabilization of homogeneous states, the heart of "self-organization," in this interplay and studied it a dozen years before anyone else even thought about it. As to the study of excitability, he proposed by the mid-1930s a deceptively simple formalism ("two-factor theory") for excitable elements, which was not substantially improved upon until the advent of Hodgkin and Huxley almost two decades later. He was also the one who conceived the pregnant idea of stringing excitable elements into networks. He thereby converted what was originally a theory of peripheral nerve into a theory of central nerve, a theory of the brain, which he bolstered by exhibiting simple networks that would discriminate, learn, remember, and do other "intelligent" things. The reader should recall that this was all accomplished in the mid-1930s. The neural network, the entire field of "artificial intelligence," and much else characteristic of contemporary research owe their very existence to Rashevsky's pioneering work of these days.

By 1950 Rashevsky had ramified his work in a hundred other directions, most of which need not concern us here. He expanded on his earlier work to successfully model a host of physiological phenomena, mostly in the area of cardiovascular and cardiopulmonary function. Suffice it to say that this modeling work was very successful, as interested readers are invited to discover, or rediscover, for themselves.

But by 1950 Rashevsky was growing uneasy. He had asked himself the basic question "What is life?" and approached it from a viewpoint tacitly as reductionistic as any of today's molecular biologists. The trouble was that, by dealing with individual functions of organisms, and capturing these aspects in separate models and formalisms, he had somehow lost the organisms themselves and could not get them back. As he himself said, some years later:

> It is important to know how pressure waves are reflected in blood vessels. It is important to know that diffusion drag forces may produce cell division. It is important to have a mathematical theory of complicated neural networks. But nothing so far in these theories indicates that the proper functioning of arteries and veins is essential for the normal course of the intracellular processes; nor does anything in those theories indicate that a complex phenomenon in the central nervous system . . . (is) tied up with metabolic processes of other cells in the organism. . . . And yet this integrated activity of the organism is probably the most essential manifestation of life.
>
> So far as the theories mentioned above are concerned, we may just as well treat, in fact *do* treat, the effects of diffusion drag forces as a peculiar diffusion problem in a rather specialized physical system, and we do treat the problems of circulation as special hydrodynamic problems. The fundamental manifestation of life mentioned above drop out from all our theories in mathematical biology.

And then he says a remarkable thing:

> . . . As we have seen, a direct application of the physical principles used in the mathematical models of biological phenomena, for the purpose of building a theory of life . . . is not likely to be fruitful. We must look for a principle which connects the different physical phenomena involved and expresses the biological unity of the organism and of the organic world as a whole.

The reader must remember that these words are coming from a mathematical physicist, who began with, and maintained to the end, the need to discover and make explicit the material basis of organic phenomena. What he was saying, on the basis of his own unparalleled experience, was this: no collection of separate *descriptions* (i.e., *models*) of organisms, however comprehensive, could be pasted together to capture the organism itself. This kind of analysis was, accordingly, not the right way to approach the organism. Some new *principle* was needed if this purpose was to be accomplished.

The search for such a principle was what Rashevsky termed *relational biology*. As usual, his words manifest an almost uncanny prescience, especially so in the light of his previous preoccupations. But in the last analysis, his main concern was with the *problem,* and as always it is the problem that tells you what you need to do.

The rich legacy of Rashevsky's work and his ideas stands in stark contrast to the fate of the man himself, a fate that I never understood but that I observed and chronicled. That fate was total neglect, if not active hostility, perhaps to be expected from experimentalists, but most especially manifested by those who should have been his natural allies, his colleagues, those who had the most to learn from him. In 1948, for example, Norbert Wiener (in his influential book *Cybernetics*) had this to say:

> Professor Rashevsky and his school of biophysics . . . has contributed much to directing the attention of the mathematically minded to the possibilities of the biological sciences, although it may seem to some of us that they are too dominated by problems of energy and potential and the methods of classical physics to do the best possible work in the study of systems like the nervous system, which are very far from being closed energetically.

This is a total, and almost certainly a willful, misconstruction of "Professor Rashevsky," especially so since these remarks occur during enthusiastic discussion of Mr. (Walter) Pitts and his neural networks, work that Pitts had done under Rashevsky's supervision and that essentially comprised a discrete, Boolean version of Rashevsky's own two-factor networks.

But at least Wiener says *something;* von Neumann had nothing to say;

Turing had nothing to say; Katchalsky had nothing to say; Prigogine had nothing to say . . . at least never in print. Richard Bellman, who at the time I first met him (ca. 1970) was in the process of redirecting his own efforts in the direction of mathematical biology, but who did not then know of my connection with Rashevsky, remarked to me that "If Rashevsky knew what he was doing, he would have been a charlatan." It turned out that Bellman had formed this opinion without benefit of ever having read any of Rashevsky's work.

Rashevsky himself, to do him credit, was indifferent to all of this, although of course he was not unaware of it. He found his happiness in his work, not in his colleagues. He was thus wiser in the ways of life, as well as of science, than many of us.

I shall not pursue Rashevsky's own approach to relational biology in what follows, although the path I do take is very much in that spirit. The reader interested in Rashevsky's own views may consult his own voluminous writings or see my earlier discussions in *FM* and *AS*.

5C. On the Concept of "Organization" in Physics

Let me begin our discussion by repeating here the key words with which Rashevsky inaugurated his investigations into relational biology:

> We must look for a principle which connects the different physical phenomena involved and expresses the biological unity of the organism and of the organic world as a whole.

Rashevsky is saying that we must look for a principle that governs the way in which physical phenomena are *organized*, a principle that governs the *organization* of phenomena, rather than the phenomena themselves. Indeed, organization is precisely what relational biology is about, as we shall see. Accordingly, it is appropriate to consider this notion, which I have not explicitly mentioned heretofore.

The term "organization" began to creep into the lexicon of physics only with the advent of the concept of *entropy* in thermodynamics, around the middle of the nineteenth century, and the attendant articulation of the Second Law of Thermodynamics. It became more prominent when reinterpreted in the course of the development of statistical mechanics. I have already discussed these matters in some detail elsewhere (see *AS*) and shall not pause to review them here.

In a nutshell, statistical mechanics allows us to *(sometimes)* interpret the classical; state variables of thermodynamics in terms of the phases of an underlying family of particles. The trick is to consider what happens on certain *sets* of phases of that underlying particulate system, rather than on individual phases, and to express this in a *probabilistic* language.

Modulo all these interpretations, a *thermodynamic state of equilibrium* becomes identified with a whole family of phases of the underlying particles. Intuitively, that family consists of all those phases for which it is in some sense "equally likely" or "equally probable" to find any particle anywhere in the system. Thus, the "state of thermodynamic equilibrium" itself becomes identified, in that sense, with a situation of total homogeneity; every piece of the system is like every other piece, insofar as the disposition (configuration) and energies (velocities) of its constituent particles is concerned. Accordingly, it seems reasonable to *identify* a "state of thermodynamic equilibrium" with a situation of *maximal disorder* or *maximal disorganization* of the constituent particles. If we do this, then by definition, every non-equilibrium state, every deviation from total homogeneity, is an *organized state,* and the extent of its organization can be measured on the one hand by its distance from equilibrium, and equivalently, by its *improbability* as a family of phases.

In this usage, "organization" becomes an attribute of thermodynamic state. It is measured at that level by its distance from the equilibrium state. Insofar as thermodynamic state can be identified with a set of underlying phases, "organization" is measured by a single number, expressing how improbable that set is.

The Second Law specifies just such a measure, at least near enough to the equilibrium state. It asserts that every *closed* system autonomously approaches an equilibrium state and that the *entropy* increases monotonically to its maximum as this happens. In more technical, formal language, entropy is asserted to be a (local) Lyapunov function for the posited autonomous approach to equilibrium. Therefore, the *value of the entropy* on a thermodynamic state measures the extent of organization (or disorganization) of that state.

The Second Law thus asserts that *a closed system cannot autonomously tend to an organized state.* Or, contrapositively, *a system autonomously tending to an organized state cannot be closed.*

This reformulation for the Second Law is suggestive, because it indicates a way of extending the notion of organization, *from state of a system, to system itself.* For if the closed *system* autonomously tends to a disorganized state of equilibrium, then "the closed system" can be thought of a setting a standard for organization (or better, for disorganization) among *systems,* just as an equilibrium state sets such a standard for states. We can therefore say that *a system is organized if it autonomously tends to an organized state.*

Hence the immediate identification of "organized system" with "open system," an identification that took an inordinately long time to be made. It took so long because, as we have seen, physics preferred to look at closed systems and tended to refer *all* behavior to closed system behavior. To this day, we talk about *organized systems* by referring their behavior back to what their behavior *would* be if they were closed, just as, in an individual closed system, we talk about an *organized state* by referring it to what it would be like at equilibrium.

These ideas comprise the full extent of the concept of "organization" in physics. These ideas have subsequently been widely exported, out of their original thermodynamic context; corresponding concepts of "organization," based entirely on *improbability,* have appeared in many areas outside of physics itself. The first conspicuous example of such an exportation was the "Information Theory" of Claude Shannon, but today there are many, many others.

For instance, there is today an important trend in the computational sciences that involves these same ideas. It proceeds from the standard identification of organization with *nonrandom,* but goes further, to identify *nonrandomness* with *entailment.* Roughly speaking, a sequence is *nonrandom* to the extent that it can be *recursively generated.* I shall describe these ideas in greater detail below, when I take up the notion of the *machine* (see chapter 8).

5D. The Concept of Function

As we saw in the preceding section, the word "organization" has been synonymized with such things as "heterogeneity" and "disequilibrium" and ultimately tied to improbability and nonrandomness. These usages claim to effectively syntacticize the concept of organization but at the cost of stripping it of most of its real content.

Indeed, when we use the term "organization" in natural language, we do not usually mean "heterogeneous," or "nonrandom," or "improbable." Such words are compatible with our normal usage perhaps but only as weak collateral attributes. This is indeed precisely why the physical concept of organization, which I have described above, has been so unhelpful; it amounts to creating an equivocation to replace "organization" by these syntactic collaterals. The interchange of these two usages is not at all a matter of replacing a vague and intuitive notion by an exact, sharply defined syntactic equivalent; it is a *mistake.*

I shall now argue that what is missing from the thermodynamic images of organization, but which lies at the heart of our intuitive usage, is tied to a concept that biologists call *function.* At first sight, this word looks at least as vague as organization itself; we shall see in a moment that this is not so.

The term *function* in biology has actually had a long and confused history. At root, it is a legacy of the machine metaphor itself. It is, however, a legacy that mechanistically inclined biologists, those most dependent on the machine metaphor, have always found embarrassing. For the term *function* has overtones of end, purpose, finality, and *telos,* ironically, just that metaphysical ideas that the metaphor was intended to do away with. Nowadays it is considered again acceptable to talk of function, because natural selection and the adaptive evolutionary processes it generates can, it is supposed, be invoked to exorcise any finalistic demons.

Nevertheless, it is relatively easy to objectify the concept of *function.* Suppose, for example, we are given a system, or better, a state, that is perceptibly heterogenous; one part looks different, or behaves differently, from another part. If we leave the system alone, some autonomous behavior will ensue. On the other hand, we can ask a question like: *if we were to remove, or change, one of these distinguishable parts, what would be the effect on that behavior?*

This is a pregnant question. It involves a new element, not merely observation, but willful, active intervention. The result of that intervention is, in effect, the creation of a *new system,* which can be regarded as a kind of perturbation or mutilation of the original one. But supposing this can be done (and I shall have much more to say about these matters as we proceed), we can compare the behaviors of these two systems, the original one, and the new one, with some original part ablated. Any discrepancy between these behaviors defines the *function* of the removed part. Indeed, as we shall see, *it provides us with another way of describing that part,* a new way of encoding that part into a formalism.

Let me introduce some neutral terminology for what I have just said. Any part of a system that can be assigned a function in the above sense will henceforth be called a *component* of the system.

From a formal point of view, the concept of *function,* and its embodiment in terms of *components,* is a part of stability theory. Namely, we are comparing two different situations: an original unperturbed one, and a second one, arising as a perturbation of the first. The discrepancy between the two *systems* defines the concept of *component;* the discrepancy between the two *behaviors* defines the *function* of the component.

This idea of component will provide for us a *unit of organization.* A part with a function, i.e., a definite relation between part and whole; this is the key that will open a whole new world of systems descriptions, indeed, a world in which organization is at the very heart.

I shall say henceforth that *any* system is *organized* to the extent that it can be analyzed into or built out of constituent components. The characteristic relationships between such constituent components, and between the components and the system as a whole, comprise a new and different approach to science itself, which we may call the *relational theory of systems.*

As we shall see abundantly, relational ideas of this type throw an entirely different light on the concept of Natural Law than is available from contemporary physics. In physics, where organization in our sense plays no role, the only entailment available is represented in terms of recursiveness of state transition sequences. On the other hand, from a relational point of view, we can meaningfully ask "why?" about organized systems in ways simply inaccessible to contemporary physics; the answers to these questions involve modes of entailment that are correspondingly inaccessible from state transition sequences alone. Thus, in particular, the formalisms (relational models) into which organized systems encode will look very different, and in fact be very different, from the ones we have exclusively seen heretofore.

5E. On the Strategy of Relational Modeling

The relational analysis of organized systems represents a most radical departure from the structural, reductionistic approach inherited from Newton. The latter approach, as we have seen, has for the past three centuries determined our view of science itself; indeed, for most of us, it simply *is* science. Though extensively modified and extended over the years, the picture first drawn in mechanics still lies at the heart of all that has followed. In the present section, I will try to indicate in the starkest manner possible the contrast between the two.

In the Newtonian approach, with its partition of the ambience into systems and environments, phases and forces, states and dynamical laws, organization as such plays no role, as we have seen. From that standpoint, it is merely an *epiphenomenon.* In the relational approach, on the other hand, the organization of a system has become the main object of study. In essence, organization has become a *thing,* with its own formal images or models, its own attributes, and its own modes of analysis.

To illustrate the vast difference between the two kinds of approach, let us see how each of them would go about telling us about an organized material system, say, a biological cell.

A Newtonian approach starts from the premise that our cell, as a material system, is to be studied and understood in the same universal terms as any other material system. That means: it must be analyzed down to a family of constituent particles. These particles define or specify a formal state space, or phase space, as we have seen; the original system, the cell, is then imaged by some special set of points *in this space.* To find the dynamical laws, we must look empirically at the different kinds of particles we have resolved our cell into; we must determine from them, in isolation, how they can interact with things around them. Specifically, we must determine both how they respond to forces imposed on them and how they impose forces on each other. From these a set of dynamical relationships (i.e., a constraint) can be written down, which specify the necessary entailments, the necessary recursions, *valid on our whole space of states or phases.* A fortiori, these entailments determine also the behaviors of our original system; these are recaptured in terms of the state transition sequences imposed by the *general* recursion rules on the *special* set of states that represent our original organized system. The only thing remaining is to mandate the initial conditions; we must *specify* one of those special states; otherwise, we will find ourselves studying some other disposition of those same particles but one that is *artifactual* as far as our original system was concerned.

In empirical terms, then, the very first step in the analysis of an organized system (e.g., our cell) is to *destroy that organization.* That is, we kill the cell, sonicate it, osmotically rupture it, or do some other drastic thing to it. We must do this to liberate the constituent particles, which are then to be further fractionated. In the Newtonian picture, this procedure has only taken us from the limited, special set of states the cell originally occupied into a larger set of states, where we have more room and can see more clearly. Moreover, in that Newtonian picture, we lose nothing by this process; once the analysis is complete, we can recapture everything about our original cell, merely by specifying *any* convenient initial state of it.

There are many things wrong with this picture. One of them, implicit in the remarks of Rashevsky that I quoted above, is this: if I give you another, different cell, then the entire analysis must be repeated for that new cell. There is nothing in the fact that the subject of analysis is a *cell* that can shorten the analysis or indeed help in any way, and when we get done, there is nothing in the resulting picture to tell us that the systems we have analyzed were cells. A consistent reductionism in biology thus converts it into a catalog or encyclopedia of individual analyses of this kind, cell by cell, organism by organism, *ad infinitum.* Indeed, people are actually engaged in trying to do this kind of thing at the present time.

In any case, I can epitomize a reductionistic approach to organization in general, and to life in particular, as follows: *throw away the organization and keep the underlying matter.*

The relational alternative to this says the exact opposite, namely: when studying an organized material system, *throw away the matter and keep the underlying organization.*

We have already, in fact, seen a version of this kind of strategy, which appears even within a completely Newtonian framework. The reader will recall my discussion of the concept of *analogy* between natural systems, as embodied in the diagram (figure 3H.3) above. We recall that two different natural systems N_1, N_2 are analogous when they realize a common formalism F. As we saw, analogy is like a modeling relation except that it relates two natural systems, rather than a natural system and a formal one.

The relation of analogy between natural systems is in fact *independent of their material constitution.* The most materially disparate natural systems can still be analogous. All that is required is that each natural system, individually and separately, be describable by the same formalism F.

Now let us suppose that the formalism F in question is *relational.* The remainder of the present volume will be devoted to making clear what that means, but for present purposes, it is enough to say that F describes a set of formal *components,* interrelated in a particular way. Any two natural systems that realize this formalism, and that are hence analogous, can thereby be said to realize, or manifest, a common *organization.* To use an old word, I can say that two natural systems N_1, N_2 that are relationally analogous share (to that extent) a common *bauplan.*

As we have seen, reductionism dispenses with organization as the first, essential step in its analysis. It expects to recapture the organization later, as I have indicated. In a relational approach, it is the matter that is dispensed with. But the concept of *realization* allows us to, in a sense, recapture matter from bauplan. Thus, for us, *realization assumes the central role,* and in particular, *the building of realizations.*

The radical departure of relational analysis from conventional analysis of material systems should now be evident. However, there is *nothing in the relational strategy that is unphysical,* in the sense of "ideal" physics. The organization of a natural system (and in particular, of a biological organism) is at least as much a part of its material reality as the specific particles that constitute it at a given time, perhaps indeed more so. As such, it can be modeled or described, in full accord with Natural Law; the resulting formalisms have at least as much right to be called images of material reality as any reductionistic model based on states and dynamical laws.

Moreover, the relational approach I have sketched in the context of natural systems has itself a formal counterpart; it is closely akin to certain aspects of generalization in mathematics (see chapter 2). As I have said, the essence of it is "throwing away the matter and keeping the organization." This means, in effect, that a relational model intentionally lacks a large number of traditional external referents; it cannot by itself be decoded into material features of a particular natural system. As we have already seen, (see, e.g., the remarks of Kleene cited in section 3D above), one of the main motivations for formalization or generalization is precisely the giving up of external referents in exchange for this presumed new generality. The only difference is that formalism attempts to pull the lost referents inside, in terms of additional syntax; in relational models this is not (and indeed, must not be) attempted. Rather, as I have indicated, we regain such referents through an additional process of *realization.*

In what follows, I shall explore the character of such relational models and formalisms, particularly with regard to their entailment structures. Then I shall turn toward the crucial concept of *realization.* In that context, I shall have to contrast relational and reductionistic formalisms *as formalisms;* it is here (see chapters 9, 10) that we will see most clearly the limitations of the latter as images of organized natural systems. It is here, too, that the concept of the *machine* will enter most directly. For the machine metaphor asserts, in our language, a relation of *analogy* between "organisms" and "machines," i.e., that there are common formalisms that they separately realize. We shall see how this assertion already fails at the formal level, when reflected to the realm of natural systems; we shall also see how that failure manifests itself in the form of *new physics.*

5F. The Component

The component may be thought of as the *particle of function;* it plays the same kind of role in relational modeling that particles play in reductionistic or Newtonian modeling. Just as in the case of particles, components for us will be the basic analytical units into which natural systems are resolved. Accordingly, I must spend time at the outset in demarcating the concept, and especially, in establishing how the component is to be endowed with formal images, how it is to be encoded into and decoded from formalisms.

From the outset, "component" is a more complicated, less intuitive concept that "particle." Some of this is simply due to the fact that we are accustomed to think in structural, particulate terms; this is part of the legacy of mechanics, which I have already discussed at some length. But the concept itself is inherently more difficult. A particle, for instance, is formally imaged in the same way, whether it is isolated, or part of a larger family; in either case, it retains all of its attributes, and hence, the *identity* that those attributes confer upon it (the situation is somewhat more complicated in quantum mechanics than in classical, owing mainly to the more abstract notion of *state* in quantum theory). But the notion of *component* is tied to that of *function,* and this in turn is dependent upon the larger system of which the component is a part. If we isolate the component, and consider it as a thing in itself, it loses its function. In other words, a functional description is *contingent* and not absolute; to describe a functional unit necessarily involves aspects outside the unit itself.

This is already an important departure from familiar ideas, which I may restate as follows: a particle, or any unit of structural analysis, does not (indeed, cannot) acquire new properties by being associated with a larger family of such units; on the contrary, the larger family is itself endowed with precisely those attributes that are contributed individually by its members (see section 6D below). Thus, a thoroughgoing reductionistic, structural approach to the natural world must deny reality to such concepts as *novelty* or *emergence* at any fundamental level. These become mere epiphenomena, new ways of collecting the same old particles. The *collection* may look new, but the particles themselves, unaffected by that fact, continue to impart their attributes to the new collection in the same old ways.

The situation is quite different with a functional unit or component. As we have seen, such a unit can by its very nature have no completely inherent, invariant description that *entails* its function; on the contrary, its description changes as the system to which it belongs changes. It can thus *acquire* new properties from the larger systems with which it is associated.

Thus, what we call a component must be endowed with the following properties: (1) it must possess enough "identity" to be considered a thing in itself, and (2) there must be enough room for it to acquire properties from larger systems to which it may belong. That is: the formal description of a component in itself as a thing; the other part must be *contingent* on such a larger system. It is this latter part that specifically pertains to the *function* of the component.

These remarks already suffice to specify how a component must be formally imaged; readers may find it instructive to stop at this point and develop the image for themselves. Before I actually do so, I will interpose a few further heuristic remarks of motivation.

Consider then the situation schematically represented in figure 5F.1.

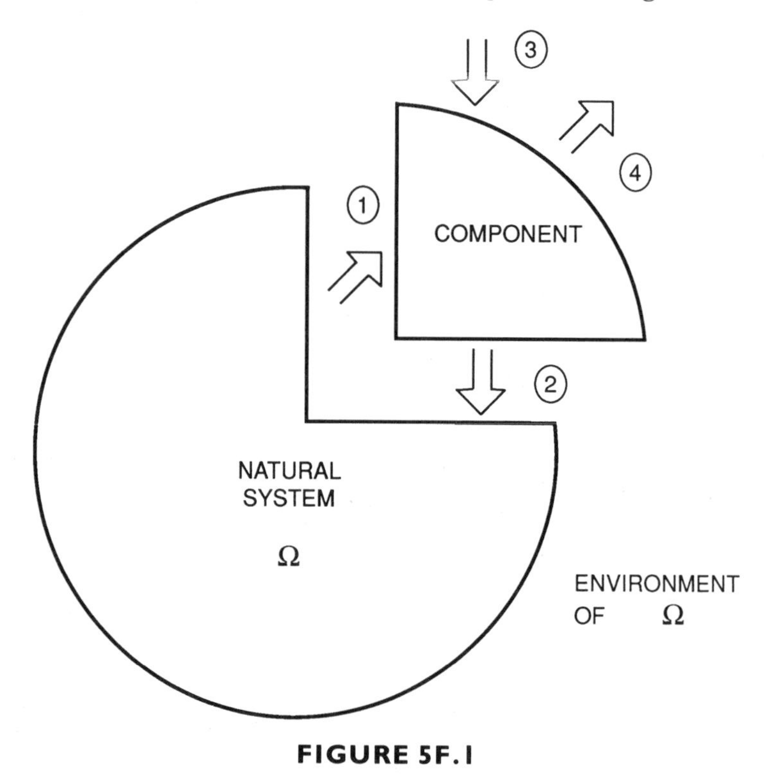

FIGURE 5F.1

In this figure, we find a *natural system* Ω, embedded in some corresponding environment. In general, this system will be *open*. Hence the environment will be the seat of a variety of influences determining the behavior of the system Ω, e.g., impressed forces, particles that can enter and leave Ω, whatever it is that sets initial conditions, etc.

In the figure, we have "lifted" a component out of Ω. Since it is a component, it plays a *functional* role with respect to Ω. This fact can be expressed by requiring the component to *interact* with the rest of Ω. That means it must (1) be affected in its own behavior by what the rest of Ω is doing, or (2) affect by its own behavior what the rest of Ω does, or both. These aspects are schematically indicated in the figure by the obvious arrows, labeled ① and ② respectively. When the component is completely removed from Ω, these arrows are cut, as indicated by the dotted lines. In addition, we must take into account the influences of the environment of Ω on the component; these are indicated schematically by the arrows ③ and ④.

Crude as it is, this diagram is suggestive in several ways. For one thing, it suggests that a component possess *an inherent polarity or asymmetry,* which we have indicated by the direction of the arrows. Specifically, there is an *input side,* or *afferent side,* reflecting the collective influence of the rest of Ω, and of the environment of Ω, on the component itself (the arrows ① and ③ in the figure). Likewise, there is an *output side,* or *efferent side,* reflecting the influence of the component (and hence its specific function) on Ω, and on the environment of Ω (the arrows ② and ④). Accordingly, to *describe* the component requires the two aspects we found necessary before: an intrinsic part (what is inside the box, labeled "component" in the figure) and a part contingent on Ω (namely, the afferent and efferent arrows, the inputs and the outputs). The function of the component depends entirely on these arrows, and hence on Ω; if Ω is changed, the arrows will change, and with it, the function of the component itself.

We have by now said enough to clearly specify what the formal image of a component must be. It must in fact be a *mapping*

$$f: A \to B$$

This formal image clearly possesses the necessary polar structure, embodied in the differentiation it imposes between the domain A of f and its range B. It also possesses the necessary duality; the "identity" of the component is embodied in the mapping f itself, while the influence of larger systems Ω, in which the component is embedded, is embodied in the specific *arguments* in A on which the mapping can operate.

In what follows, I shall never use the term "function" in its mathematical sense, as a synonym for "mapping"; I reserve it entirely as an expression of the relation of components to systems and to each other.

As the reader might expect, there are many consequences of passing to this new kind of encoding of material reality, some of them of the most profound import. I shall take these up in due course, as I proceed with our discussion.

5G. Systems From Components

I have already suggested that a component be thought of as a "particle of function." We have also seen that the original development of Newtonian particle mechanics was a *synthetic* theory, building from the behaviors of individual particles to families of particles. When this synthetic theory was combined with the atomic hypothesis, that *every* material system could be *analyzed* down to such a family, particle mechanics became universal, with the consequences I have described above.

It is clear that a corresponding synthetic relational theory of systems is incipient in the encoding of *components* into abstract *mappings*. For instance, we have at our disposal the formal concept of *composition of mappings*. In simplest terms, whenever the domain of one mapping intersects the range of another, the mappings may be *composed*, applied in succession to yield a composite mapping. Formally, if $f : A \rightarrow B$, $g : C \rightarrow D$ are mappings, and if $B \cap C \neq \emptyset$, then a *composite map* $h := gf$ may be defined for all a in A such that $f(a)$ is in $B \cap C$.

The simplest synthetic situation is thus the juxtaposition of two suitable components in a simple *series circuit*. Specifically, if $f : A \rightarrow B$, $g : C \rightarrow D$ are components (or more precisely, represent components), and if $B \cap C \neq \emptyset$, then we can form the *composite*, as in the diagram below:

$$A \xrightarrow{f} (B \cap C) \xrightarrow{g} D. \qquad [5G.1]$$

This composite represents our first example of a larger (formal) relational system, containing the given components as parts, and with respect to which the parts acquire specific functions.

Let us note that the composite diagrammed in [5G.1] can itself be thought of as a *component*. Hence, in particular, it itself may be used synthetically and incorporated as a part of a still larger relational system. Indeed, if we define $\varphi = gf$, we can represent [5G.1] by the elementary diagram

$$\varphi : A \rightarrow D. \qquad [5G.2]$$

Thus we can see that the *analysis* of relational systems is associated with the *factorization of mappings;* conversely, the synthesis of relational systems is associated with the *composition* of mappings.

Let us look at a slightly more complicated kind of relational system that we can synthesize from components. Suppose we are given the components (mappings)

$$\begin{aligned} f &: A \rightarrow B \\ g &: B \rightarrow C \\ h &: B \rightarrow D \end{aligned}$$

Then we can construct the composite system represented by the diagram

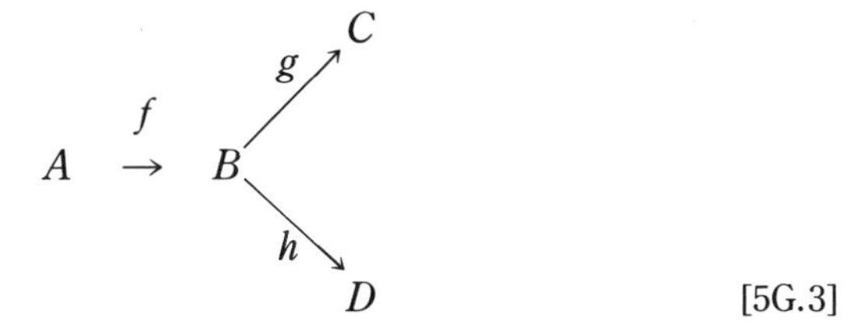

[5G.3]

in which B is a "branch point." Here again, I have synthesized a larger system [5G.3] from given components. Moreover, just as before, we can collapse this branched diagram back down to a single component, with the aid of standard formal constructions. Indeed, if we write

$$\varphi : A \rightarrow C \times D \qquad [5G.4]$$

where φ is defined by

$$\varphi(a) = (gf(a), hf(a))$$

for every a in A, we see that [5G.3] and [5G.4] are formally equivalent.

Slightly more complicated still is the diagram [5G.5]:

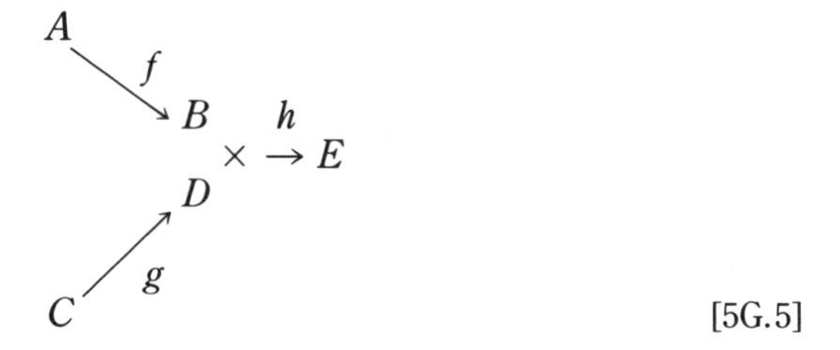

[5G.5]

which we have synthesized from three mappings

$$\begin{aligned} f &: A \rightarrow B, \\ g &: C \rightarrow D, \\ h &: B \times D \rightarrow E. \end{aligned}$$

This last map h is a perfectly good mapping, but its domain happens to be expressible as a Cartesian product of other sets. We can see from this simple situation how such Cartesian products serve as "confluence points" in synthetic relational diagrams.

With the help of Cartesian product operations, on sets and on mappings, we can once again formally express [5G.5] as a single mapping, of the form

$$\psi : A \times C \rightarrow E$$

where ψ is defined by

$$\psi(a, c) = (h(f \times g)(a, c). \qquad [5G.6]$$

for every *a* in *A*, and every *c* in *C*.

We leave it to the readers to satisfy themselves that arbitrarily complicated diagrams of mappings can be constructed in this way. Each of them represents a possible relational system; following my own earlier usage, I shall call such a diagram an *abstract block diagram* of the relational system to which it corresponds, and whose organization it describes. It is clear that a relational system may have many equivalent abstract block diagrams, because at this level, we are completely free to amalgamate mappings into composites, and conversely, to express given mappings as factorizations of new ones. We shall soon see, however, that this freedom is not as great, or as arbitrary, as it presently appears to be.

In these terms, then, we can now be a little clearer about the term "organization." Namely: *organization is that attribute of a natural system which codes into the form of an abstract block diagram.* To enlarge on our earlier image: if the component is an *atom* of organization, then a general abstract block diagram is the *molecule* of organization.

Thus, for us, organization is itself a rather complicated concept; it involves a family of sets, a corresponding family of mappings defined on these sets, and above all, the abstract block diagram that interrelates them, that gives them *functions.* We have seen that the formal operations of composition and factorization of mappings provide great flexibility in the generation of abstract block diagrams. But we can also see, from a consideration of the few examples given above, that if we simplify the abstract block diagram by using fewer mappings (i.e., if the mappings we use are composites, and hence further factorizable), we pay for this by requiring more complicated sets on which these mappings act. We shall see later that, in a certain sense, the "degree of organization" of a system is *independent* of any specific abstract block diagram we use to represent it, being spread in equal parts over the sets, the mappings, and the diagram itself.

Thus, as promised, organization has become a *thing.* We have seen in particular how the thing that organization is could be characterized, represented, and studied in the abstract, i.e., divorced from every other aspect of material nature. Specifically, no reference has been made to particles, or states, or forces, to any of those things that heretofore have exclusively constituted our armory of attack on material nature. Of course, we must ultimately come to terms with these, if for no other reason than that they constitute the conventional language of science today and that our experimental tools (particularly in biology) are geared to see matter rather than functions or relations. But it is already clear, I hope, what was meant by asserting that the relational strategy proceeds by *throwing away the physics and retaining the organization.* It is also clear, I hope, that a relational description, in the form of an abstract block diagram, is no less an expression of material reality than a description built from states and dynamical laws. Indeed, I shall come to argue that it is much more so.

5H. Entailments in Relational Systems

In the preceding sections, I have made a gentle and natural progression from *organization* to *function* to *component,* and thence back to organization (now expressed explicitly in terms of relations between components). In the process, I argued that the formal image of an organization is an abstract block diagram of mappings. In the present section, I shall begin our exploration of the consequences of these ideas. The present discussion will be relatively unsurprising but will set the stage for the far more radical conclusions that follow.

I begin with a consideration of the formalisms involved in relational modeling of organized systems. As we have seen, the formal world into which we have been ushered is a world of sets and mappings. There is a name for such a world, which we have seen before in another context; that name is *category.* As we shall see, it is no accident that the concept of category arises of itself in these several, apparently very different contexts.

As I have repeatedly stated, any formal system is specified in terms of the entailment structure that characterizes it. Categories, as formal systems, happen to be preternaturally rich in entailments. For one thing, there are the entailments that characterize a category as a simple mathematical object, just as, e.g., the group axioms characterize a group as a mathematical object. But the *contents* of a category are themselves rich in entailments. For a category is inherently a *dualistic* structure, consisting of two different *kinds* of things: *sets* and *mappings.* For now, we will think of this dualism as absolute; the far more radical developments that follow involve what happens when we drop this unnecessary stipulation.

An individual mapping in a category,

$$f : A \to B,$$

is in itself already a vehicle of entailment. It says specifically that, for any element a in its domain A,

$$a \Rightarrow f(a)$$

(to be read as "a entails $f(a)$"). It is more accurate to write that

$$f \Rightarrow (a \Rightarrow f(a)). \qquad [5H.1]$$

This usage makes it explicit that the mapping f, and the specific argument a on which it acts, are each required to entail the image $f(a)$; it also exhibits that these modes of entailment are *different.* I discuss this further in a moment.

Since there are generally many mappings in a category, there are correspondingly many modes of entailment within the category itself; accordingly I shall call such entailments *inner entailments.* The dualism between sets and mappings in a category can be expressed (or, better, expresses itself) in terms of these modes of entailment, in the following way:

1. Only the *elements* of sets can be entailed within a category; neither the sets in the category, nor the mappings between them, are entailed by inner entailments.
2. If an element b of a set B is in fact entailed, then it must lie in the *range* of a mapping f.
3. In that case, we can write $b = f(a)$, for some element a in the *domain* of f.

The relations between b, a, and f, which express these conditions, are given precisely by [5H.1] above.

It is instructive to consider the significance of these inner entailments for the *components* they encode. We recall that according to our formulation of Natural Law, entailment *within* a formalism corresponds to *causality* in that part of the ambience described by the formalism. As we have seen, a component encodes into a mapping; the inputs to the component encode into the elements of the domain of the mapping; the outputs of the component encode into the elements of its range. Accordingly, if we think of an element $b = f(a)$ as an, effect (that is, if we ask the question "why $f(a)$?"), then we can answer: "$f(a)$ because a"; and also, "$f(a)$ because f"; see section 3 above. In the Aristotelian parlance, it is clear that the first answer posits *input as a material cause of output $f(a)$;* the second answer posits *mapping of as efficient cause of output $f(a)$.* I defer discussing the relational image of formal cause, and of final cause, until a bit later. But it is already clear that causal entailment in relational systems is a very different thing from what we have seen before; we see now no recursion, no chronicles, indeed, no such thing as time.

So far, we have considered only inner entailments, existing within a category simply by virtue of the mappings it contains, and the sets on which these mappings operate. As we have seen, these inner entailments are of a restricted kind; only *elements of sets* can be entailed thereby. This suffices for the encoding of individual *components,* as we have just noted. But in order to pass from components to abstract block diagrams, we must invoke other entailment structure, structure that comes from the defining properties of categories themselves. As I shall now indicate, these *outer entailments* allow us to entail (in fact, to *construct*) mappings from mappings in the category, and sets from sets.

Since these matters will soon become crucial for us, and since there are some small subtleties involved, we will consider them in detail. We begin by looking at the simple series circuit

$$A \xrightarrow{f} B \xrightarrow{g} C \qquad [5H.2]$$

which we have seen before, and which is perhaps the simplest abstract block diagram. We have two individual mappings f, g; each of them determines an entailment rule [5H.1]. Namely, we have for the mappings f, g the entailments

$$f \Rightarrow (a \Rightarrow f(a)),\ \text{all } a \text{ in } A;$$
$$g \Rightarrow (b \Rightarrow g(b)),\ \text{all } b \text{ in } B;$$

If it should happen that we can put

$$b = f(a)$$

in the above, then the second rule gives

$$g \Rightarrow (f(a) \Rightarrow g(f(a))).$$

What we want to do, obviously, is to replace the expression $g(f(a))$ by a new one, $(gf)(a)$, which satisfies

$$(gf) \Rightarrow (a \Rightarrow (gf)(a)). \qquad [5H.3]$$

Or in other words, we want the individual mappings f, g to entail a new mapping, which we have written as (gf), such that

$$(gf) : A \rightarrow C.$$

Clearly, we cannot accomplish such a thing with inner entailments. Rather, we must invoke a *category axiom,* the one that allows *composition of map-*

pings. In effect, this axiom gives us an inferential rule F of precisely the form

$$F(f, g) = gf. \tag{5H.4}$$

F is basically a binary operation, whose domain consists of pairs of mappings in the category, and whose range is mappings. Via [5H.4], the composite gf is now explicitly *entailed* by its factors. To use some suggestive language, gf can be regarded as an *effect;* the pair (f, g) is its material cause, and the inferential rule F is its efficient cause. Thus we can now write, analogous to [5H.1] above,

$$F \Rightarrow ((f, g) \Rightarrow gf). \tag{5H.5}$$

But it is crucial to keep in mind that this inferential rule F, which governs the entailment of mappings from mappings, is not *in* the category; *it does not encode a component.* What it actually does turn out to encode will soon become very interesting to us.

If we now turn to the abstract block diagram exhibited in [5G.5] above, we notice another kind of outer entailment that must be invoked. This one governs the entailment of a Cartesian product from its individual factors, i.e., the entailment of a set in the category from other sets. Indeed, it specifies a binary operation on *sets,* just as F specified a binary operation (composition) on *mappings.* It can accordingly be described in exactly the same terms as I have used to describe F; the detailed discussion will be left to the reader.

We can formally do a great deal with the modes of inner and outer entailment inherent in any category. In particular, we can concatenate them to form, and characterize, arbitrarily complicated abstract block diagrams from the sets and mappings in any particular category. In fact, the totality of abstract block diagrams that can be formed in this way constitutes a new category, entailed in the obvious sense by a given category C. Indeed, this category $D(C)$ stands roughly in the same relation to C as a (free) monoid A^{-}s stands to its set A of generators. or, to use another metaphor I introduced earlier, if the category C is thought of as comprising the "atoms" of organization, then $D(C)$ comprises the "molecules." One could obviously spend a great deal of time studying these ideas in the abstract. But this is not our primary interest; the real novelty of a relational ideas lies a bit deeper.

51. Finalistic Entailment: Function and Finality

In the preceding section, we have become acquainted with some of the modes of entailment available in relational formalisms. We could already distinguish two kinds of entailment, which we called *inner entailment* and *outer entailment;* the former were embodied in individual mappings the formal images of components), while the latter came from the global inferential rules governing the totality of components. As we saw, the inner entailments allow us to entail elements; the outer entailments allow us, in a limited sense, to entail mappings and sets as well.

In this formal universe, the term *function* seems to have dropped from explicit sight. It is, of course, present; it is embodied directly in the notion of component itself and in the relations between components in an abstract block diagram. In the present section, we shall explore these relationships, an the functions they embody, in terms of entailment. In particular, we want to explore the question of whether, and how, a component *entails* its functions in such a diagram.

As I noted earlier, the notion of function in biology was extrapolated from experience with machines. A machine is built from *parts,* according to a specific design. The parts of the machine play definite roles in determining the activity of the machine as a whole; they have, or have acquired, specific *functions* by virtue of their inclusion into the machine. Indeed, these acquired functions can be thought of as *part of the design;* we can say that the parts of a machine are there precisely for the sake of the functions they execute in the activity of the machine as a whole.

The machine metaphor suggests very strongly that the "parts" of organisms are analogous to the parts of machines. Hence they are there for the sake of whatever functions they execute with respect to the organism as a whole. The machine metaphor does not actually *say* this; indeed, it *cannot* say this because such a stipulation would require an accurate specification of what a machine is. This in turn was far beyond Descartes and those who followed him; as we shall soon see (see chapter 7), it is still not an easy question.

In any case, if the parts of a machine are there by design, and if the functions of those parts are part of the design, what does that say about organisms? Obviously, nothing very good. For by invoking the concept of design, and the explanation of parts in terms of design (i.e., in terms of the *functions* manifested by parts), we are talking about finality. This is all right when we talk about machines as human fabrications, but it is manifestly not all right to consider organisms in such terms.

The central issue for biology here is: how can we have organization without finality? Nowadays, biologists generally believe they have papered over this issue. In a nutshell, Darwinian evolution through natural selection, with its attendant adaptations, serves precisely to do this. The argument is that the produce of an evolutionary process gives the *appearance* of design but without any of the finalistic implications of design. Through evolution, then, we can have organic machines, in which parts have functions, but shaped entirely by natural selection and not by fabrication. At least, that is the claim.

As such, the *explanation* of function then devolves upon the evolutionary process itself, and not upon the particular relation of part to whole that process has generated. It thus remains, strictly speaking, illegal to explain, e.g., the *function* of mitochondria (i.e., to answer the question of "why mitochondria?") by referring to the exigencies of energy generation; this is only a *façon de parler,* a shorthand for a whole evolutionary chronicle, and never to be taken literally.

On the other hand, it is clear that biology can never rest content with this mode of explanation alone. Indeed, the bulk of biology itself is not at all concerned with elucidation of evolutionary chronicles, but rather with trying to understand organisms as individual natural systems in their own right. What, indeed, is reductionism about, if not the analysis of organic behaviors, per se, into convenient units, in terms of which they may be explained and understood? And as we have seen, reductionism is merely the biological manifestation of a universal strategy of analysis of natural systems, totally *independent* of whatever evolutionary history those systems may have had.

The whole thrust of the old Aristotelian analysis of causation is to make it manifest that no one mode of causal entailment suffices to understand anything. At root, this is because *the causal categories do not entail each other.* In terms of machines, for instance, we may know how to build a watch, without knowing anything about how a watch actually works; conversely, we may know how a watch works, without any idea of how to build one. In terms of biology, we may know all about evolution, but nothing about physiology, and conversely. In other words: the modes of entailment involved in evolutionary processes, and the modes of entailment involved in physiological processes, are *different.* They are not unrelated, but they are not equivalent or interchangeable. And in particular, any exclusion of finality from evolution does not thereby exclude it from physiology, or conversely. I shall return to this matter in more detail subsequently, when I explicitly turn to questions of *fabrication.*

Let us return to the consideration of entailments within organizations per se, specifically, organizations embodied in abstract block diagrams. We are concerned now with entailments associated with function, entailments imposed by component on system, and those imposed by system on component. It is precisely here that finality manifests itself.

We have already noted that final causation has been a difficult concept because in a certain sense, it proceeds in the opposite direction from the other causal categories; (see section 3E above). Namely, if we specify something as an *effect,* and hence ask "why?" about it, then the several answers that specify its material causation, efficient causation, and formal causation, all in their various ways *entail* the effect. Final causation is different; a final cause of an effect is defined in terms of *something entailed by the effect.* All the troubles occasioned by the concept of finality spring from this curious fact.

For instance, it is standard to say that the final cause of a house is an intention to have a place to live. But "a place to live" involves things entailed *by* the house, things not entailed by nonhouses. Thus it is completely equivalent to say "a house is entailed, as effect, of its final cause, by an intention to reside in it" and to say "a house itself entails further effects, some of which are synonymous with residence therein." To make finality respectable, it suffices to replace the former usage by the latter, to look at what is entailed *by* an effect.

But this in turn means that an effect must have something nontrivial to entail. Now we can understand why finality is so resolutely excluded from Newtonian encodings. First, as we have seen, entailment in that picture is embodied entirely in the recursiveness of state transition sequences. There is nothing in that picture for a state to entail except a subsequent state. Furthermore, a state can itself be entailed only by a preceding state. The presence of *time* as a parameter for state transition sequences translates into an assertion that *causes must not anticipate effects.* Therefore, whether we express final causation in terms of "intentionality," or equivalently in terms of what its effect entails, final causation in the Newtonian picture involves the *future acting on the present.* And of course, this is clearly inconsistent with the encoding of the other causal categories *in the Newtonian picture.*

This is a basic point, so let us recapitulate. In the Newtonian picture, a *state* can only entail *subsequent states.* (That is all the entailment present in the Newtonian encoding, as we have seen). Subsequent states are necessarily later in time than present states. Finality is expressible only in terms of *what is entailed by* a state, and hence, in the Newtonian picture, *only* in terms of future states. *Ergo,* final causation, as a separate causal category, cannot exist in that picture.

In the relational picture, on the other hand, the situation is quite different. As I have developed it so far, there is no time parameter, no states, no state transition sequences. There are only components (mappings), and the organizations, the abstract block diagrams, which can be built from them. In any such diagram, there is plenty for a component to entail. Indeed, what it entails in an abstract block diagram is precisely what I have called its *function* in the diagram. In this situation, then, it is perfectly respectable to talk about a category of final causation and to identify a component as the effect of its final cause, its function in the diagram. In this sense, then, *a component is entailed by its function,* in any particular abstract block diagram in which it appears.

It may seem at first that we gain little from this terminology, aside from legitimizing its usage in biology. But in fact, we gain exactly what Aristotle said we do, *another independent mode of entailment.* Moreover, it is a mode that *entails a mapping,* in an entirely different way than the entailments we have seen heretofore. To have a name for it, I shall call it *functional entailment,* distinct from the inner and outer entailments that were previously introduced. All of them will, individually and collectively, play crucial roles in what follows.

There is nothing unphysical about functional entailment. What is true is that functional entailment has no encoding into any formalism of contemporary physics; it represents a notion of final causation that is *unencodable* in any such formalism from the outset. On the other hand, it reflects basic features of material organization per se. At root, it is the resolute exclusion of these features, these manifestations of matter, that makes contemporary physical formalisms so special. Put baldly, there is simply not enough entailment in these formalisms to encompass biology. But that is a fault in the formalisms, and in the encodings into them; it certainly does not connote anything vitalistic or transphysical in material nature. And above all, it is entirely consistent with Natural Law.

5J. Augmented Abstract Block Diagrams

In the past few sections, we have seen some of the modes of entailment available to us in abstract block diagrams and in the categories that are their habitats. They already provide the basis for a very rich entailment structure, much richer than anything available from recursion alone. However, they are by no means the only entailments available in categories, nor are they for our later purposes the most important. In the present section, I take up yet another class of entailments, a class that will play a preeminent role in what follows.

Everything I have discussed so far involves an absolute separation between *sets* (the domains and ranges of mappings) and *mappings* (which encode the components themselves). So far, then, it is the elements of sets that can be entailed in abstract block diagrams, by making a set the range of a mapping, the output of a component. There is in particular *no way to entail a mapping,* no way to have a mapping encode an *effect.* Mappings must either be given or else built out of given mappings through relatively trivial applications of outer entailments.

But there is nothing in Category Theory that mandates this absolute distinction between sets and mappings. Quite the contrary: indeed, if A and B are arbitrary sets, then the totality $H(A, B)$ of mappings from A to B is again a *set.* There is nothing to prevent $H(A, B)$ from being a set in any category to which A and B themselves belong. As such, $H(A, B)$ can be the range of other mappings in the category and therefore *entailed* by precisely the same inner entailments that we have already seen.

In other words, it is perfectly consistent with Category Theory for mappings f to encode *effects,* and hence, give answers to questions of the form "why f?" Heretofore, we had no way of doing this; mappings could only *entail* effects, i.e., answer questions of the form "why $f(a)$?" Now, however, they themselves can be entailed, by other inner entailments within the category. It is precisely here that the real power of relational ideas becomes manifest, and the most radical departure from Newtonian ideas occurs.

The basic situation we are envisaging is a diagram of the form

$$\varphi : X \rightarrow H(Y, Z). \qquad [5J.1]$$

Here, X, Y, Z are arbitrary sets in the category, and $H(Y, Z)$ denotes a set of mappings (components) from Y to Z. Hence φ itself encodes a component, which for any x in X, entails an image $\varphi(x)$ in the set $H(Y, Z)$. So far, all is as it was before. The novelty resides in the fact that $\varphi(x)$ is itself a *mapping;* in fact, given any y in $Y, \varphi(x)$ entails a new element

$$(\varphi(x))(y) \text{ in } Z.$$

This kind of situation is not comprehended within the abstract block diagrams we have built so far. The relation between φ and any of its images $\varphi(x)$ is in fact of a kind we have not seen before; hence we need a new notation to express it. We shall write

$$X \xrightarrow{\varphi} H(Y, Z)$$
$$\Big\downarrow \text{(dotted)}$$
$$Y \xrightarrow{\varphi(x)} Z \qquad [5J.2]$$

where the dotted arrow indicates that $\varphi(x)$, as a mapping from Y to Z, is itself *an output of* φ; more precisely, that there is an x in X that entails, via φ, the mapping $\varphi(x)$ from Y to Z.

We can now build a much larger class of abstract block diagrams, which contain such dotted arrows. I will call these *augmented block diagrams.* I stress that, in these augmented block diagrams, the dotted arrows must not be confused with composition of mappings. In simple composition, the output of one component is the input to another; the range of one mapping lies in the domain of another. Such a situation is always represented by a pair of solid arrows. The dotted arrows, which indicate that a component *is* an output of another, cannot be composed; they denote an entirely new and different relation between components that we have seen heretofore.

Of particular interest for us is the special class of [5J.2] in which we put

$$X = Z.$$

Then [5J.2] takes on a characteristic form, which we will write as

$$Y \xrightarrow{\varphi(z)} Z \xrightarrow{\varphi} H(Y, Z). \qquad [5J.3]$$

(with a dotted arrow curving from $H(Y, Z)$ back to $\varphi(z)$)

In what is to follow, we shall see quite a lot indeed of this diagram and its correlates.

For the moment, I mention one important feature of this particular augmented abstract block diagram. In a sense, it represents the *most organized* relational system we have yet seen; it contains fewer sets and more arrows (entailments) than any I have exhibited so far. Indeed, under the conditions I have developed so far, it is *maximally organized;* it requires us to independently posit the smallest number of independent, unentailed elements.

Let us now proceed a step further. So far, we have only considered situations in which a set of the form $H(X, Y)$ can be the *range* of another mapping, i.e., in which a component can be an output of another. The dual situation obviously comprises components of the form

$$\psi : H(X, Y) \rightarrow Z, \qquad [5J.4]$$

in which components can be *inputs* to others.

And now, of course, the final step, in which mappings can be both inputs and outputs to other mappings, i.e., in which both domain and range of a mapping are themselves sets of mappings. That is, we consider diagrams of the form

$$\theta : H(X, Y) \rightarrow H(U, V). \qquad [5J.5]$$

These last will be very important to us.

For instance, let us consider once again the augmented abstract block diagram [5J.3]. We notice that there is still an unentailed mapping in that diagram, namely, the mapping $\varphi : Z \rightarrow H(Y,Z)$. Suppose we want to entail it. As we know, in relational terms, this means essentially putting φ into the range of some new mapping, and hence, enlarging the diagram. Well, we certainly know one set to which φ belongs, namely, the set of mappings $H(Z, H(Y,Z))$. Hence one way of entailing φ is to find a mapping β whose range is this set and adding it to the diagram.

The simplest thing to do is to form the diagram [5J.6]:

$$Y \xrightarrow{\varphi(z)} Z \xrightarrow{\varphi} H(Y, Z) \xrightarrow{\beta} H(Z, H(Y, Z)) \qquad [5J.6]$$

That is, we suppose that the *domain* of β is itself a set of mappings or components, of the form [5J.5]. In this way, we entail the mapping φ, at the cost of introducing a new unentailed mapping β. However, the diagram [5J.6] is more organized than that of [5J.3], in the sense that the ratio of unentailed elements in the former is smaller than in the latter.

It is evident that the diagrams [5J.3] and [5J.6] represent only the first two steps of a process that can be iterated indefinitely. At no finite stage is everything in the diagram entailed; the last mapping at any stage in this sequence is always itself unentailed. Nevertheless, the diagrams themselves are becoming more and more organized as we progress through the sequence, in the specific sense that the ratio of unentailed elements to entailed elements is approaching zero. In this sense, in the "limit," everything would be entailed.

These few comments should remind readers of several things we have seen before, in quite different contexts—for instance the Gödel Incompleteness Theorem, which says that no finite syntactic entailment scheme can entail everything and stay consistent, and also the states and recursion in terms of Taylor's Theorem, in which an infinite number of independent temporal derivatives were needed to specify the recursive object (the state). In the former case, however, nothing can be done to break off the infinities. In the latter case, as we saw, there are situations in which the infinities do break off, and the recursive object (state) can be characterized in finite terms, namely, in the presence of *constraints*, identical relations between the temporal derivatives.

In very rough terms, the formal system whose augmented abstract block diagram is the infinite iteration of [5J.3] and [5J.6] is analogous to the infinite recursive object in Taylor's Theorem. We can likewise ask whether that infinite diagram can be made to break off, in the same way that an identity among successive temporal derivatives makes the state specification break off. This, as we recall, was precisely the upshot of Newton's Second Law, which allowed us to replace an infinite dimensional state by a finite dimensional phase. In our relational situation, the analog of the Second Law would require a way of *entailing one of the mappings in our sequence from the sets and mappings that precede it in the sequence.* For instance, if we could somehow make the map φ in [5J.6] already be entailed by what precedes it in the diagram, the result would be a closed, maximally organized formalism in which *everything is already entailed.*

The question we have just raised has been couched entirely in terms of entailment within formalisms. Specifically, it has to do with how one must go about acquiring the most entailment, and hence the most organization, from a family of sets and mappings between them. As such, it is at this level a purely formal question; indeed, as we presented it, it is a part of Category Theory. We shall soon meet, however, with this identical question in an entirely different set of circumstances. I defer answering it until that time. When we do meet it again, however, in chapters 8 and 9, it will become the focal point of our entire enterprise.

5K. Finality in Augmented Abstract Block Diagrams

Whenever we introduce new modes of entailment into a formal system, we thereby endow the elements of the system with new *functions.* Consider our augmented abstract block diagrams in functional terms, and hence, in terms of Aristotelian ideas of finality.

As we saw earlier, a simple component

$$f: A \rightarrow B$$

could be described entirely in terms of the entailments

$$f \Rightarrow (a \Rightarrow \mathrm{f}(a)).$$

In Aristotelian language, $f(a)$ is the *effect;* its argument a corresponds to *material cause,* and f itself corresponds to *efficient cause,* of that effect. But $f(a)$ can have nothing corresponding to *final cause.* This is because, as we recall, final cause of an effect is tied to what is entailed *by* the effect, and there is obviously nothing for $f(a)$ to entail.

In other words, the question "why $f(a)$?" in this situation can only have the two answers: "because f", and "because a." On the other hand, the questions "why f?" and "why a?" in this diagram can be answered *only* in finalistic terms. The mapping f, and its argument a, are both unentailed in the diagram, and hence, cannot themselves be effects, and therefore do not have material or efficient or formal cause. They do, however, *entail* something in the diagram, namely the effect $f(a)$. This means they have *function* (namely, to entail $f(a)$), and their function legitimately answers the questions "why?" about them. Indeed, we see that it is the *only* legitimate answer to the question available in these circumstances.

Summing up, then, the causal correlates of the simple diagram $f : A \rightarrow B$, we have the following:

1. If we treat the image $f(a)$ as *effect,* then $f(a)$ has material and efficient cause but no final cause;
2. If we try to treat the map f and its argument a as *effect,* then they have *only* final cause, identified with what they entail (namely, with $f(a)$).

Note explicitly that final cause is not a unique mode of entailment, in the same sense that the other causal categories are. For instance, if we have a different mapping

$$g : C \rightarrow B,$$

and pick an element c in C such that

$$g(c) = f(a), \qquad [5K.1]$$

then the questions "why g?" "why c?" receive exactly the same answers as before, even though $g \neq f$ and $c \neq a$. There can thus be many distinct, different ways of entailing the effect $f(a)$ (i.e., of *executing* the function), between which final causation cannot distinguish. That is, the mappings f and g, and their respective arguments a and c, are *equivalent* in terms of function. This equivalence, or similarity, here manifests itself precisely through the invariance relation [5K.1].

Thus it is that finality is allied to the notion of *possibility,* while the other causal categories involve *necessity.* This fact constitutes another reason for the unpopularity of finality among those conversant only with a Newtonian tradition in science. For it is inherent in the brief discussion I have just given that (1) the same *function* can be manifested in many different ways, and also, conversely, (2) many *different* functions are incipient in any relational unit. We may put these facts another way: there is nothing about a component per se that entails any particular function it may manifest, nor is there anything about a particular organization that entails a specific component. This is not, at root, a cost of the relational strategy of replacing matter by organization.

As we shall see abundantly later, the real power of relational analysis in general, and of relational biology in particular, lies precisely here. Indeed, it is precisely by treating organization as a thing in the abstract that we are enabled to see the arbitrary (i.e., unentailed) nature of the relation between an organization and the components that are its elements. That is, there is nothing in the components that mandates that particular organization, nor anything in the organization that mandates those particular components. It is only after an organization has been specified, by means of positing a definite abstract block diagram, that its components acquire specific functions, and the resulting entailments *within* that organization can be analyzed and explored. All the rest lies in the specific character of particular *realizations,* whose particularities are not entailed relationally.

Of course, such ideas do not sit well with those who analyze only individual material *realizations* of particular modes of organization. In any such particular situation, the organization is "wired in" to the physics of the system from the outset. This in turn conveys a strong, but utterly mistaken, impression that the physics is entailing that wiring, rather than (in a precise sense) the reverse. This impression encourages them to look for entailments from material structure to functional organization, entailments that are not in fact there. Nevertheless, the temptation is strong to attempt to argue "from structure to function," to posit *unique* relationships between the two, and to dismiss denials of such relationships as "holism." To them, what they call "holism" is sterile, precisely because it denies these entailments, on which their activities depend. But the relational approach provides new, different entailments of its own, as we shall soon see abundantly.

In fact, I remark in passing that the basis for a *comparison* of organization (i.e., for establishing a precise notion of *analogy* between abstract block diagrams) can be built entirely from [5K.1] above. I shall not pause to do this here, since it would take us too far afield at the present moment. I remark, as a hint for the interested reader, that analogy between abstract block diagrams means the encoding and decoding of one diagram into another in a structure-preserving way; hence it involves functors between diagrams and between categories of diagrams. It was in this direction, in fact, that Rashevsky was moving in his initial pioneering work on relational biology and that was embodied informally in his *Principle of Biotopological Mapping.* But that is another story.

Let us return, after these tangential remarks, to a consideration of finality in abstract block diagrams. I have so far spoken of finality only with respect to components

$$f : A \to B. \quad [5K.2]$$

As I pointed out, the images $f(a)$ have as yet no function, because they have nothing to entail. On the other hand, f itself, and its arguments, have only function; they are themselves unentailed.

By embedding a component in a larger abstract block diagram, we can endow the images $f(a)$ of the component with something to entail, and at the same time, entail the arguments a, and even the mapping f, from other sets and mappings in the abstract block diagram. That is, we endow the images $f(a)$ with particular *functions,* while other components in the diagram *acquire new functions* with respect to f itself.

Clearly, the more entailment present in such an abstract block diagram, the more functions are acquired by a component f with respect to the system as a whole, and conversely, the more functions are acquired by the other components of the system with respect to f. We shall restrict ourselves to looking at one particular situation: namely, one we have already specified as maximally entailed. This is the augmented abstract block diagram exhibited in [5J.3] above, which I now write as

$$A \xrightarrow{f} B \xrightarrow{\Phi} H(A, B). \quad [5K.3]$$

(with a dashed arrow curving from $H(A, B)$ back to f)

We have now given $f(a)$ something to entail, for each argument a. We shall be particularly interested in the case for which

$$\Phi \Rightarrow (f(a) \Rightarrow f) \quad [5K.4]$$

or what is the same thing

$$\Phi(f(a)) = f. \quad [5K.5]$$

The new ingredients in this diagram are the mapping $\Phi : B \Rightarrow H(A, B)$, and the augmented, dotted arrow, here indicating that Φ has entailed the original mapping f. We have:

1. The new mapping Φ, which is itself unentailed. If we ask the question "why Φ?" its only answer in the diagram is "because f." Hence, as with any otherwise unentailed unit of a formalism, the only answer to this question is in terms of function and finality.
2. The original mapping f, which in the new situation is now entailed. Thus, if we ask the question, "why f?" we can answer it now as follows:
   i. "because $f(a)$"
   This was, of course, the original *finalistic* answer, which was all that was available when f was unentailed. This answer remains unchanged in the new diagram. But we now also have:
   ii. "because $f(a)$"
   This says that f is entailed by its *value* $f(a)$. More precisely, from [5K.5], this value is analogous to *material cause* of f itself. And also,
   iii. "because Φ"
   Once again, [5K.4] says that f *is* the value of Φ at the argument $f(a)$, and hence Φ is analogous to *efficient cause* of f.

So in the new diagram, we have three answers to the question "why f?" where we had only one before. But if one looks at these answers, we find a most curious circumstance: *(i) and (ii) are the same.* Final cause and material cause are identical. The *function* of f in the augmented diagram, which objectively looks the same as it did before, *has now become part of the entailment of f.*

Because of the loop that is closed by the dotted arrow in [5K.3], then, we now have the unique circumstance that *the function of f is the entailment of f.*

This merging of function (final cause) with entailment (material cause) clearly arises whenever we have a loop in an augmented abstract block diagram. Furthermore, such loops are themselves the consequence of putting enough entailment (i.e., enough organization) into a formalism. It is thus a generic correlate of the organizational level of a relational system, i.e., of the ratio of unentailed elements to entailed ones.

Clearly, the circumstances I have just indicated cannot arise in an unaugmented abstract block diagram, in which none of the mappings are entailed. It also cannot happen in a Newtonian framework, where "finality" is tied up with the future, and entailment with the past. As we shall soon see, these considerations, which so far are of purely formal, are in fact intimately bound up with life itself. They are invisible from any but a relational standpoint, because the very basis for them does not exist in the scientific epistemology we have inherited. From this perspective, we can begin to really see why the fundamental question, "what is life?" has proved so refractory.

5L. The Theory of Categories

I have already had many occasions to mention the Theory of Categories in the course of our discussion, and we shall see much more of it as we continue. It may be helpful to digress and discuss it in some detail. What is in many ways most instructive and illuminating about it is its history, which illustrates better than anything else how mathematics models itself and how it seeks to generalize by a process of forgetting about the system of referents that produced the model in the first place.

Thus, I begin with a brief description of this history, not with the standard formal definitions, which anyone can read in a host of available references. For the moment, then, I am concerned with the historical scaffolding that culiminated in the theory; then we can turn to some of the salient features of the theory itself.

The Theory of Categories arose out of topology. As we have already noted, present-day topology is itself the confluence of several initially distinct historical streams. One of them is embodied in the Cauchy concept of convergence, which involved the passage to limits of sequences of numbers. As we saw, along with a notion of convergence comes a notion of continuity in mappings defined on numbers. Any mapping pulls a sequence in its domain into an image sequence in its range; the mapping is continuous if convergence of the domain sequence entails that of the range sequence. In other words, a mapping is continuous if it commutes with the operation of taking limits of sequences; if the image of a limit is the limit of the images.

The quest for the most general (i.e., the most impoverished) situation in which convergence *and* continuity can be discussed has culminated in what today is called point-set topology. After a long period of experimentation, which did not end in consensus until well into the present century, that most general situation is now embodied in the concept of a *topological space;* this is a composite structure involving a set of points and a system of subsets ("neighborhoods", or open sets) satisfying a few basic conditions.

On the other hand, another essential ancestor of present-day topology comes from the concerns of geometry. On the face of it, geometry seems to be concerned with quite different questions, e.g., with geometrical figures and the relations between them. Here, the key notion (according to the Erlangen Program of Klein) is that of congruence and the allowed motions that bring congruent figures into coincidence.

We can already see difficulties in trying to put these two circles of ideas together. Namely, convergence and continuity are local concepts; they involve what happens in the vicinity of specific points. On the other hand, the traditional figures of the geometer are not local objects. Interplay between local and global is always hard; that is in fact one thrust of this entire volume, and in part is why the history of Category Theory, which arose out of precisely such an interplay, is so instructive.

The obvious way to connect the two, in the present case, is to tie the notion of congruence (a global notion) to that of continuity (a local one). Klein's Program leaves us free to specify the group of motions that define the geometry ad lib, subject only to the proviso that they form a group. This is, however, a strong restriction; most continuous mappings are not invertible. Thus, if we are to combine the two ideas, we need continuous mappings (to preserve the local structure) that are also invertible (so as to form a group). In a word, we are led to the *homeomorphisms,* which turn a topological space into a geometry in the sense of Klein. Thus, in a topological space, two subsets ("figures") are congruent if and only if there is a homeomorphism that brings them into coincidence. In the process of doing this, of course, we establish thereby an explicit homeomorphism between the figures themselves.

If the topological space in question is already a familiar geometric object, like the Euclidean plane, or the surface of a sphere, a geometry based on homeomorphism is in a sense the most general kind of geometry we would want to consider. It is most general in the sense that more kinds of figures become congruent under the group of homeomorphisms than under any other. Hence the traditional characterization of this situation as "rubber-sheet geometry."

Now we can observe two things. First, the concept of homeomorphism between two topological spaces does not require them to be subsets of a common space that contains them both. It seems to make sense entirely in the abstract. If we wish to retain the original geometric flavor of the Klein Program and treat individual topological spaces as "figures," we can see ourselves being led toward a *category* of all topological spaces. In the category, then, each constituent space plays the role of a single geometric figure, to be compared with others.

There is thus the problem of trying to decide whether two particular topological spaces are homeomorphic or not. This is the Classification Problem, the central problem in all of topology. Homeomorphism has to be an equivalence relation between topological spaces; thus, solving this Classification Problem means the characterization of the corresponding equivalence classes. The Theory of Categories arose historically in the context of trying to do this.

The Classification Problem is, of course, a global kind of problem. One way to approach it is, as with any equivalence relation, in terms of *invariants,* with numbers or other objects that are constant on the equivalence classes. In the present case, this means associating with each topological space a corresponding number (or a more general object) in such a way that homeomorphic spaces get the same number; thus two spaces that are assigned different numbers cannot be homeomorphic. The trouble is that nonhomeomorphic spaces may also get the same number, so we only have in general a necessary condition. The hope is that by finding enough invariants (i.e., a complete set) we can turn all these necessary conditions into one big sufficient one as well.

Most of what is presently available in these directions goes back to Poincaré. In a sense, Poincaré's basic strategy was to retreat to local considerations again, not local in the sense of the neighborhoods of point-set topology but in the sense of going back to small pieces or subsets, into which whole spaces could be consistently atomized. These ideas were originally embodied in what came to be called combinatorial topology, nowadays algebraic topology. In our terms, it embodies a very syntactical approach to topology (or more specifically, to the search for "enough" topological invariants to solve the Classification Problem).

Let us briefly review Poincaré's ideas, because they are interesting, and because they shed light on the historical origins of Category Theory. Basically, Poincaré started from the observation that, if we take two points close enough together inside a space, they might determine (bound) a line segment also lying entirely in the space. Call such points zero-simplexes; the line they bound a one-simplex. Several such one-simplexes (a one-chain) may bound a two-simplex, e.g., a triangle, again entirely within the space. Iterating, several such two-simplexes may bound a volume (a three-simplex), and so on. We thus obtain, at each stage, a simplex of a certain dimension, which is a chain of simplexes of lower dimension. We also have a natural idea of boundary, on which the whole approach is based. By doing this carefully, the entire original space is replaced by (or approximated by) an array of such simplexes (a complex). The idea of chain, and boundary, can be readily extended from simplex to complex.

By judiciously assigning *orientations* to simplexes, we can extend the notion of chain so that, for each dimension n a simplex may have, the resulting generalized chains form an abelian group. These chains may be thought of as routings for paths one may traverse in a complex along simplexes of that dimension. The boundary operation serves precisely to relate these groups of chains in a very elegant way. It turns out that the boundary operation, appropriately generalized, is a homomorphism between groups of chains of successive dimensions. In particular, at each dimension n, the boundary operation isolates two particular subgroups: (1) those that are the images (boundaries) of chains of one higher dimension, and (2) those that themselves are bound by chains of one lower dimension. The latter is always a subgroup of the former, so we may form the quotient group; call it H_n. This is the nth homology group of the complex; its elements are certain equivalence classes of chains. The details can be found in any text on algebraic topology. The importance of H_n is that it is a topological invariant of the complex, and in fact, of the topological space that the complex approximates, at least, under circumstances where this notion of "approximation" can be made explicit enough.

Thus, with each complex S, which is itself a topological space (plus a lot of additional structure), we assign a family of groups $H_n(S)$, in a topologically invariant way. Moreover, if S_1, S_2 are two such complexes, and $f: S_1 \to S_2$ is a continuous mapping between them, there is a corresponding group homomorphism

$$H_n(f) : H_n(S_1) :\to H_n(S_2)$$

for each n; if f is a homeomorphism, then $H_n(f)$ is an isomorphism. From these facts, we can conclude that S_1 and S_2 can be homeomorphic *only if* $H_n(S_1)$, $H_n(S_2)$ are isomorphic as groups for each n. However, we cannot in general conclude the converse; the isomorphism of $H_n(S_1)$ and $H_n(S_2)$, as groups, does not generally imply that S_1 and S_2 are homeomorphic. Thus the procedure above gives us *some* invariants but *not enough* to solve the Classification Problem. Moreover, it works at all only if we can give a sense to the replacement of a topological space by approximating complexes.

These H_n are examples of *functors*. They basically associate groups $H_n(S)$ with complexes S, and group homomorphisms to continuous mappings, as I have indicated. It is most important to observe that such factors do not, in general, associate or encode *elements* (points) of S with *elements* of $H_n(S)$. By perusing the steps of the argument, the reader will see that we lost the elements of S (and with them, the details of their local neighborhood structure) at an early stage, in fact, in the passage from a complex to the groups of chains. In fact, all we retain about S is that some of its points lie on boundaries that separate others; i.e., a *relational* property.

Indeed, the relation between S and the $H_n(S)$ is what we earlier termed (see section 3I above) a *metaphorical* one. That is, we can decode from the $H_n(S)$ something about S; we can decode from the algebraic relations between $H_n(S_1)$ and $H_n(S_2)$ something about the topological relations between S_1 and S_2. And it is important to note that the metaphor (in this case, from the $H_n(S)$ to S, from groups to spaces) runs in the opposite direction as the functor itself (i.e., from spaces to groups). That is, we decode back from the *range* of the functor to its *domain.*

Category Theory itself embodies this picture but divorces it completely from any specific referents. Thus, in place of the topological spaces and groups of algebraic topology, we have unspecified *objects;* in place of continuous mappings and group homomorphisms, we have correspondingly unspecified *morphisms.* In fact, we can even dispense with the objects and proceed entirely on the basis of morphisms alone. We require only that morphisms can be *composed,* just as mappings can, and that this composition is associative. It is also convenient to mandate that every object in such an abstract category comes with an identity morphism. A functor then becomes simply a rule for associating objects with objects, and morphisms with morphisms, which preserves these identities and the composition of morphisms.

But we are not yet done. If we think of functorial images as constituting metaphors for what they image, we may ask whether two different such metaphors are the same, or if not, whether one in some sense includes the other. That is, we can compare metaphors, compare functors, by attempting to build, in effect, modeling relations between them. This is where modeling explicitly enters the picture, not (yet) between the domain and the range of a single functor, but between the ranges of different functors defined on a common domain. This kind of comparison is accomplished by a *natural transformation,* as I now proceed to describe. In the historical development of Category Theory, according to its inventors, the concept of natural transformation came first, and the rest of Category Theory came as the underpinning for it.

If A and B are categories, a functor

$$T : A \rightarrow B$$

turns a diagram of the form

$$A_1 \xrightarrow{f} A_2$$

where A_1 and A_2 are objects in the category A, and f denotes a morphism between them, into a diagram of the form

$$T(A_1) \xrightarrow{T(f)} T(A_2)$$

in the category B. If T_1, T_2 are two different functors, we get two distinct diagrams in B, of the same *form* but built out of different constituents in general, namely,

$$T_1(A_1) \xrightarrow{T_1(f)} T_1(A_2)$$

and

$$T_2(A_1) \xrightarrow{T_2(f)} T_2(A_2)$$

Now the objects A_1, A_2, and their functorial images, lie in different categories in general (here, in A and B respectively), so there is no question of their being related directly by morphisms. But those functorial images themselves now lie in the *same* category (i.e., in B); hence they may be related in B, by morphisms belonging to B.

Specifically, suppose we can find morphisms

$$\theta_1 : T_1(A_1) \rightarrow T_2(A_1),$$
$$\theta_2 : T_1(A_2) \rightarrow T_2(A_2),$$

such that the entire diagram

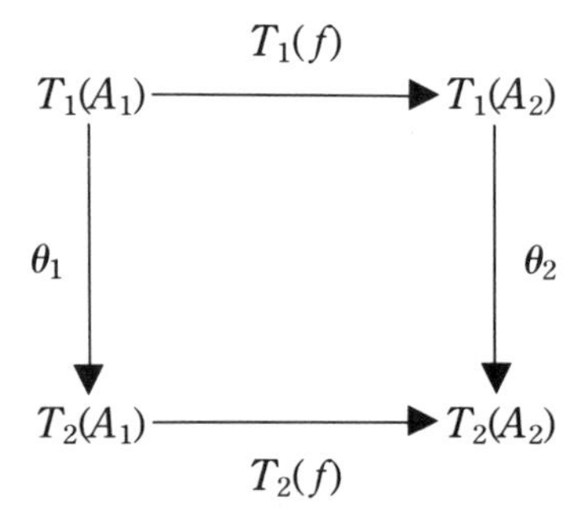

commutes. Then we have in effect *encoded* the functorial images of T_1 thereby into corresponding functorial images of T_2; i.e., we can embed the metaphor T_1 *into* the metaphor T_2. In that sense, we can *model* T_1 in T_2.

If we can find such a morphism θ_A in B, for every object A in A, then the totality $\mathrm{H} = \{\theta_A\}$ of them constitutes a *natural transformation*

$$\mathrm{H} : T_1 \rightarrow T_2$$

If the morphisms θ_A are invertible for each A in A, then the resulting natural transformation is called a *natural equivalence* between the functors, for obvious reasons.

This is where, in my view, the real power of Category Theory ultimately lies. It is impossible in this short space to fully substantiate this last remark, but let us give two brief indications of why I assert it. The first arises from the very form

$$\mathrm{H} : T_1 \rightarrow T_2$$

in which we have expressed the relation between two functors T_1, T_2 and a natural transformation H between them; H looks quite like just a morphism, with T_1, T_2 constituting its domain and range respectively. This is in fact correct; we can take functors as objects of a new category, with natural transformations providing the morphisms between them. And we can keep doing this, iterating this procedure, so that the functors and natural transformations arising at any given level become the objects and morphisms for the next. In a certain sense, then, Category Theory can talk about itself, or describe itself, in ways more nearly akin to natural languages than to the formal systems that normally constitute mathematics.

The second remark is the following. In the discussion above, we have tacitly supposed that A and B are different categories; that functors $T : A \rightarrow B$ allow us to make metaphors in B for what is happening in A and that natural transformations allow us to model such a metaphor in another. However, if we put $A = B$, or more generally, if we can put A and B into a common category, then natural transformations can serve to turn metaphors into models. In particular, if $T : A \rightarrow A$ is a functor, and if T is a natural transform of the identity functor, then $T(A)$ is actually a model of A, for each A in A.

Let us illustrate these ideas with a little example. It was already shown by Eilenberg and MacLane, in an appendix to their historic paper of 1945, that any (small) abstract category A could be embedded into, or realized by, a subcategory of the familiar category ß of sets. They did this by actually constructing a functor $R : \mathsf{A} \rightarrow$ ß, which accomplishes the required embedding. Let us run through their argument, which is extremely short.

We want to associate each object A of A with a set $R(A)$. Put $R(A)$ equal to the *set* of all morphisms $f : X \rightarrow A$ in A with range A; that these constitute a set is precisely what "small" means. Moreover, if $\varphi : A \rightarrow B$ is a morphism in A, let us define

$$R(\varphi) : R(A) \rightarrow R(B)$$

by

$$R(\varphi)f = f\varphi.$$

It is readily verified that R is a functor and that it does the trick.

Now observe that this construction works, even if A is already a subcategory of ß i.e., even if its objects are already sets, and its morphisms ordinary mappings between them. In this case, however, it makes sense to ask whether R is a natural transformation of the identity. We would then be able to find mappings θ_A, θ_B such that the diagram

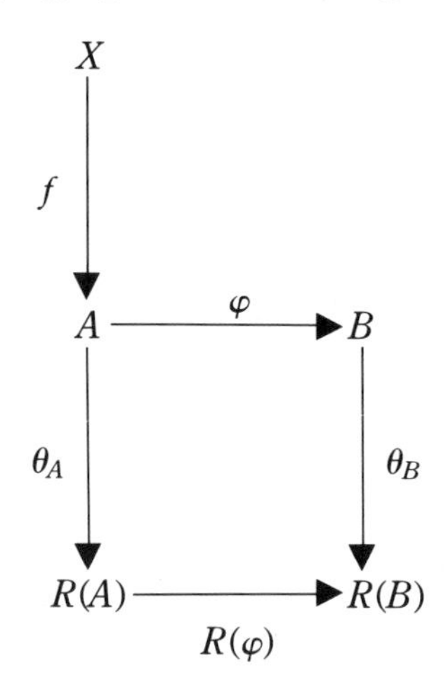

commutes. But what is of present interest is the left column of the diagram; i.e., the composition

$$X \xrightarrow{f} A \xrightarrow{\theta_A} P(A).$$

But this is a diagram we have seen before; it is in fact [5J.3] in a different notation. We thus can see how a diagram of this kind, what we called an augmented block diagram, can arise in a very general context, one intimately connected with the concept of model itself, as manifested in the Category-theoretic idea of a natural transformation.

With a little care (actually, a lot of care), the functor R can be iterated. That is, it can be applied twice in succession, to define a new functor R^2. We can then ask whether R^2 is, in turn, a natural transform of the identity. If it is, the diagram above can be further augmented. These remarks should be kept in mind; we will see them again in section 10C below.

This concludes our little survey of the Theory of Categories. I have tried to indicate how it arose from remote roots in arithmetic and geometry, through successive processes of generalizations and confluences. The generalizations in question were all made by temporarily giving up specific referents, referents that seemed essential at the time to the very generation of the ideas being generalized but that could be discarded, the ideas themselves separated from them, and finally, in this form, realized in entirely new ways than that with which we started. As we have seen, this is in itself a very relational thing to do. The end result in this case, Category Theory, is quite unique among mathematical formalisms; it comprises within itself a general theory of modelling per se and indeed a flexibility remarkably approaching that of natural language.

I do not believe that anyone has yet fathomed all the things that Category Theory can do, even after decades of work. It can, of course, be employed as a tool, for talking coherently about specific referents. It can be studied as a thing in itself. It can be used as a kind of transducer, to move ideas and methods from one part of mathematics to another. But each of these, important as they may be, is only a part, and preoccupation with a part obscures the whole.

Chapter 6

Analytic and Synthetic Models

IN THE preceding chapters, we have made the acquaintance of most of the main characters, which will animate the remainder of our considerations. Another, called *simulation,* still waits in the wings, awaiting its cue; this will come in the next chapter. For the present, I will consolidate what we already have and introduce some basic ideas needed to set the stage for its ultimate appearance.

Modeling, as we have abundantly seen, is the art of bringing entailment structures into congruence. It is indeed an art, just as surely as poetry, music, and painting are. Indeed, the essence of art is that, at root, it rests on the unentailed, on the intuitive leap. I have stressed repeatedly that the encodings and decodings on which modeling relations depend are themselves unentailed. Thus, theoretical scientists must be more artists than craftsmen; Natural Law assures them only that their art is not in vain, but it in itself provides not the slightest clue how to go about it.

6A. Modeling Relations

Modeling relations can be thought of as transductions, in which one kind of entailment can be replaced by, or converted into, another, in an invariant way. We convert causal entailments to inferential ones for a very simple and basic reason; we can hope to *understand* what goes on in a formal system. The concrete manifestation of such understanding lies mainly in our capacity to *predict* and (though this is a quite different order of question) perhaps to control.

It would be nice if we could pull the modeling process itself inside a formal system, where we can see it whole. We cannot do this directly, but we can do it metaphorically, by looking at what it takes to bring two *formal* systems in congruence. That is, we can contemplate the modeling of formal, mathematical objects by means of other formal objects. As we have described at length elsewhere (see *AS*), there is indeed a long and very interesting history of modeling within mathematics itself, which is very beautiful and very illuminating, and to which I have already alluded several times; see section 3F above. We thus have at hand a collection, a museum in the true sense of the term, in which epistemological and ontological problems do not exist. The system to be modeled is no longer the proverbial "black box"; we can see inside it to an extent that would be Godlike in the external world. In particular, we can see, and articulate, *relations between models*, and relations between *realizations of models*, in a detail simply not available to us in scientific modeling.

This is the task of the present chapter. It is of basic importance to us because I have already introduced two entirely different kinds of models of natural systems, the Newtonian and the relational, which we need to interrelate and to understand if we are going to do biology. One way to prepare ourselves for this is to explore the mathematical metaphors for modeling itself, where both model and modeled are themselves formal systems.

In the course of this investigation into relations between models of a given formal system, and between realizations of a given model, we shall again meet the Theory of Categories. As we have seen (see section 5I above), this is itself a very complicated kind of formal system, which was originally built to exemplify experience with modeling in a particular formal domain (algebraic topology) but is in fact much richer than that. It is, among other things, the natural habitat for discussing, not only the specific modeling relations that have historically arisen in mathematics, but also the enterprise of modeling itself. It is thus unique among formal systems in its inherent reflexive characteristics, which as I said, approach those of natural language, unique to the point that many mathematicians, even today, do not consider it to be mathematics at all. Nevertheless, we shall meet it in many different guises as we proceed.

As we shall see, an examination of modeling within mathematics reveals two quite different approaches to modeling. One of them, which I shall call *analytic*, is intimately tied to the idea of direct (Cartesian) product; in the context of natural systems, this approach is tied to the notion of efficient cause. The other, which I shall call *synthetic*, is tied rather to the idea of *direct sum* and to the notion of material cause. In the Theory of Categories, these two ideas turn out to be dual to each other in a precise sense, but they are generally very different. Only in very special situations do they happen to be equivalent in any sense, and these special situations all inherently involve some kind of *linearity*. Nevertheless, it will turn out that the modeling enterprise itself, both in science and mathematics, has been tacitly predicated on the coincidence of the analytic and the synthetic approaches, i.e., on the equivalence of direct products and direct sums.

In fact, the difference between direct sum and direct product, between synthetic and analytic models, is also closely allied to the difference between *syntactic* and *semantic.* In this light, we can already see the profound effects of supposing from the outset, however tacitly, that they coincide. Indeed, from this viewpoint, Gödel's Incompleteness Theorem is an assertion that direct product $\neq$ direct sum (more specifically, direct product is "bigger," in a sense to be made precise below). In the chapters to follow, we shall see further that an assumption of coincidence between analysis and synthesis leads directly to the concept of *mechanism* (via the auxiliary concept of *simulation,* which will be developed in due course; see section 7D et seq. below).

In a sense, it is the thrust of this entire work that this hypothesis of analysis = synthesis must be dropped. Above all, it must be dropped if we are to do biology, and hence a fortiori, it must be dropped if we are even to do physics. By dropping it, we enter a new realm of system, which I call *complex,* and which in certain sense needs to have no synthetic models at all. The distinction between relational and Newtonian models of natural systems will become crucial here, because as we shall see, the former extend to the realm of complex systems, while the latter cannot.

6B. Some Preliminaries: Equivalence Relations

One of the most primitive concepts in all of mathematics is that of an equivalence relation on a set. We have already seen such relations several times above, e.g., section 4D above. Since it is basic to our concerns in many ways, not least through its relation to the direct or Cartesian product, I shall begin with this.

At one level, equivalence relations can be looked upon as a vehicle for constructing new sets (quotient sets) from given ones. Specifically, given an equivalence relation R on a set S, we construct a new set S/R, the set of all equivalence classes of S under the relation R. But in another sense, the relation R does more than this. It allows us to decompose the specification of any individual point s in S into two steps: first, find in S/R the equivalence class to which s belongs, and then, locate s within this class.

A property of s that depends only on its class (i.e., on what it projects to in S/R) will be called a *general property* of s, relative to R. On the other hand, a property of s that depends on where it is within its class (i.e., on

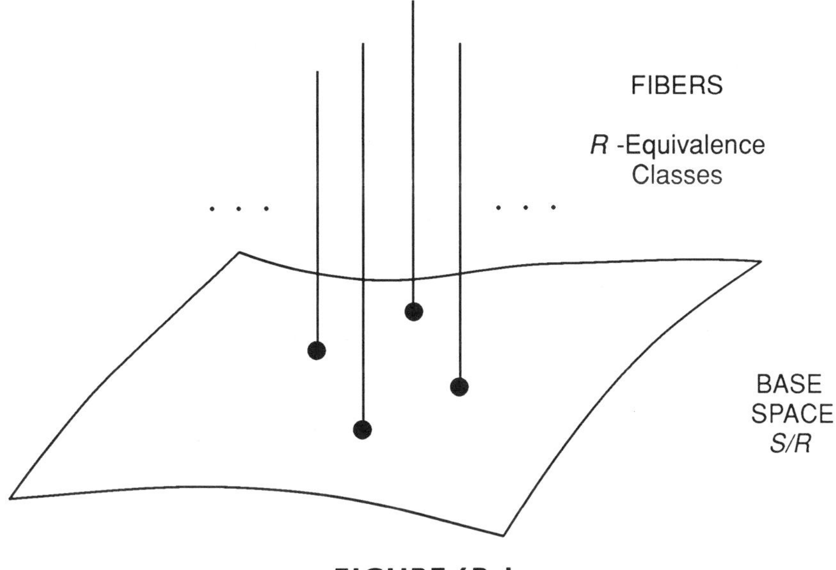

FIGURE 6B.1

what *distinguishes* it from other points in its class) will be called a *special property.* The equivalence relation R itself clearly segregates, and recognizes, only the general properties. To see special properties, we need to step outside the equivalence relation; we need to assume the Godlike perspective to which I alluded previously. (The reader should compare this usage of the terms *general* and *special* with the discussion in chapter 2 above.)

A general property may be called an *R-invariant.* For instance, any mapping $f : S/R \to X$, where X is any set, can be uniquely "lifted" to S itself. This "lifting" has the obvious property that it is constant on the equivalence classes of R; its values depend only on general properties, not on special ones. Such a mapping is an R-invariant.

It may be helpful to introduce a geometric visualization for the partition into special and general arising from an equivalence relation R. We can think of the set S/R of equivalence classes as a *base space.* Over each point in this space, there is the equivalence class, the *fiber* consisting of all elements of S that project onto this point. We thus have a picture like figure 6B.1.

At this level of generality, the fibers are in general all different from one another. Clearly, the base space here is the habitat of the general properties under R; the fibers embody the special properties. Accordingly, we can regard any point of S as comprising a pair (a, x), where the first "coordinate" pertains to the equivalence class of the point (i.e., embodies what is general about the point with respect to R), and the second "coordinate" embodies what is special.

Let us look at the two extreme cases of this kind of situation. First, if equivalence is actually equality, then S/R and S coincide. With respect to equality, then, every property is general, none are special. At the other extreme, there is the trivial relation that says that all points of S are equivalent. Here, every property is special. But a vestige of generality remains, namely, the elementary fact that S itself constitutes a set *that we can name.* In a certain sense, this trivial relation retains, or recognizes, only the property of "belonging to S", i.e., of elementhood.

There is in fact a very close relationship between naming and the concept of equivalence. Indeed, if $f: S \to X$ is any mapping, then we know f induces an equivalence relation R_f on S, by

$$s \, R_f \, s' \text{ if } f(s) = f(s').$$

That is, two points of S are equivalent if and only if the mapping f cannot distinguish them, if and only if f assigns them the same image, *the same name.* We then have the obvious identification

$$S/R_f \simeq f(S)$$

which constitutes an encoding of the individual elements of S into the set X. Clearly, the general properties of the elements $s \in S$ under R_f are those that f can recognize or distinguish; the special properties are those that are invisible to f. Conversely, because S/R_f and $f(S)$ are isomorphic, we can *decode* from $f(S)$ to S/R_f, but not in general back to S itself; only those properties of the elements of S that are *general* can be decoded.

Let me now introduce some useful terminology. I will henceforth call the mapping $f: S \to X$ an *observable* of S; the totality of values $f(S)$ of f will be called its *spectrum.* It is conventional to take $X = R$, the set of real numbers, and I shall henceforth do so (though it is not necessary and in many situations is actually too restrictive). In any case, we can see that the evaluation of a real-valued mapping on a point of its domain is a formal metaphor for the measurement process, by which an observer "gains information" about a natural system he or she is contemplating. The information he or she gains is embodied in a specific value, the outcome of the measurement, which serves to name or label or encode something about the system itself. These ideas provide the basis for discussion in *FM*, which should be consulted for further details.

I cannot emphasize too strongly that, in the formal world, S is already a determinate entity (in this case, a set), so that in general, looking at S as imaged in the spectrum $f(S)$ of an observable inevitably "*loses* information" about S. In the case of a natural system, on the other hand, the counterpart of S is initially *unknown,* veiled completely in its noumenal and phenomenal shrouds. The whole purpose of measurement in this case is to *provide information* about it.

Let us pause for a moment here and scrutinize this innocent-looking phrase, "provide information"; it involves a subtlety we shall see again many times. In terms of our discussion of sections 3D and 3E above, we can identify the term "information" with "answer to a question." That is, we have tied the concept of "information" to a linguistic form (the interrogative), which exists only in natural language, and not in formalisms. So what are the questions in this case that turn $f(s)$, the value of f on s, into information about s? One of them, certainly, is "why $f(s)$?" to which one answer is "because s." Thus, we can think of s itself as something that can answer the question "why $f(s)$?"; in our present situation, there is a whole equivalence class full of such answers, which f cannot distinguish.

But there is an important presupposition, one built into the very notation we are using. When we write the expression "$f(s)$," we get the impression that f, the mapping, is fixed, and s, its argument, is variable in S. But the role of argument and mapping are *formally* interchangeable; we could just as well keep s fixed in S, and let f, the mapping or observable, vary in a set of such. Then formally, s itself becomes a mapping, whose arguments are themselves mappings; we would then write $s(f)$, or $\hat{s}(f)$, where

$$\hat{s}(f) = f(s).$$

This is, in fact, a familiar mathematical trick, which looks innocuous but is in fact of great moment. We can see this from the fact that we can answer the question "why $f(s)$?" with the answer "because s," just as correctly as with "because $\hat{s}$." Clearly, in these two answers, the causal status of s and $\hat{s}$ are completely different, even though s and $\hat{s}$ are formally identifiable, different names for the same thing. (The reader should here recall the discussion of section 5J above, on the entailment of mappings.)

Intuitively, if we think of the observable or mapping f as corresponding to a *meter,* and the argument s of f as something the meter is measuring, it is more natural to think of $\hat{s}$ rather than s. For the measurement process results in a *change in the meter* (e.g., a deviation of a dial from a reference position); it is in fact this change or deviation that embodies the measurement, which we *impute* to what is measured and which concretely embodies the information conveyed by the measurement. Seen in this light, the answer "because s" to the question "why $f(s)$?" pertains to efficient cause of the effect $f(s)$. This in fact the interpretation we shall ultimately want.

These remarks should be kept in mind as we proceed. I also call attention to the fact, exemplified by these remarks, that apparently innocuous changes in *interpretations* (encodings and decodings) can have the most far-reaching conceptual significance.

In any case, the formal question "can we retrieve S itself from the spectra of observables?" is the metaphoric formal counterpart of perhaps the basic question of epistemology. Accordingly, let us look at it for a moment.

In the rich formal world of set theory, when a set S is given, many other sets are determined as well. We have the Cartesian products $S \times S$, $S \times S \times S, \ldots$; we have the set 2^S of all subsets of S; we have the set of mappings (observables) $H(S, X)$, where X is any other set (we will, of course, be specifically interested in $H(S, \mathbf{R})$); we have all the conventional set-theoretic operations (unions, intersections, complementations, inclusions). The question I have asked, namely, "can we retrieve S from the spectra of its observables?" can thus be approached with a formidable battery of inferential structure.

For instance, given S, the set $R(S)$ of all equivalence relations on S is determined. It is in fact a certain set of subsets of $2^{S \times S}$. The usual operations of union, intersection, and complementation are available in $R(S)$; we can talk about the union of two equivalence relations; their intersection, etc. We also have a natural partial ordering (inclusion); we can say that one equivalence relation is bigger or smaller than another. In fact, then, $R(S)$ has an enormous amount of internal structure; it is a *lattice.*

Note that, in formal terms, $R(S)$ is itself an encoding of S itself into another set $R(S)$. It is *not* an encoding of the *points of* S but of S as a whole. The symbol R here can be thought of as representing itself a kind of observable, whose arguments are now *sets,* and whose spectrum is a set of sets. It is in fact an example of a *functor,* which at this level operates, not on individual systems, but on whole collections *(categories)* of them. Seen in this light, the question we are asking (namely, can we retrieve S from the spectra of observables on S?) is a *category-theoretic question;* it involves something about S as a whole, something meaningless at the level of the individual elements of S. And the answer to this question pertains not *just* to S specifically, but to *every* S in a corresponding category.

So let us return to that question, with these ideas in mind. From our godlike perspective, we know there is a close relation between $H(S, \mathbf{R})$, the set of all observables, and $R(S)$, the set of all equivalence relations on S. In fact, if f is any observable, we know we can associate with it a definite equivalence relation R_f in $R(S)$. These are the ones we are interested in.

Intuitively, if one observable f tells us something about the elements of

S, then two observables (f, g) should tell us more. Specifically, each element s of S now gets *two* labels $f(s)$, $g(s)$ respectively, two encodings, two names. Formally, we can define a new equivalence relation

$$R_{fg} = R_f \cap R_g$$

on S from R_f and R_g. What are its equivalence classes? Obviously, if s, s' are two elements of S, then

$$s\, R_{fg}\, s' \quad \text{if} \quad \begin{cases} f(s) = f(s') \\ \quad \text{and} \\ g(s) = g(s'). \end{cases}$$

That is, two elements s, s' are equivalent under R_{fg} if neither f nor g can tell them apart. The equivalence classes of R_{fg} are thus the *intersections* of all the equivalence classes of R_f with all the equivalence classes of R_g.

In other words: R_{fg} has *more classes* than either R_f or R_g alone. The classes themselves are thus smaller. Under R_{fg}, then, more properties of the points of S are general properties; fewer are special. In formal terms, it is said that R_{fg} is a *refinement* of R_f and of R_g. Hence the encoding

$$s \rightarrow (f(s), g(s))$$

tells us "more" about s than either of the constituent encodings

$$s \rightarrow f(s), \; s \rightarrow g(s)$$

did separately.

We note explicitly that there need not (though there may) already be an observable $h : S \rightarrow \boldsymbol{R}$ such that $R_{fg} = R_h$. If so, then clearly R_{fg} and R_h are naturally equivalent. Moreover, refinement itself constitutes an instance of natural transformation.

In causal terms, two observables f, g provide more "information" than either one alone because we are now asking two independent questions, namely, "why $f(s)$?" and "why $g(s)$?" and proposing the single answer "because s" to *both* of them. As pointed out earlier, it would be more precise to propose the answer "because $\hat{s}$" to these questions, but the profound difference between these formally equivalent answers will become important to us only later.

We can iterate this process in the obvious way. Take a *family* $\{f_1, \ldots, f_n, \ldots\}$ of observables and form the equivalence relation

$$\bigcap_i R_{f_i}.$$

Clearly, this encodes an element s of S into the corresponding family of values

$$s \rightarrow (f_1(s), f_2(s), \ldots, f_n(s), \ldots).$$

The question is whether we can find such a family for which the resulting equivalence is actually ***equality.*** In more formal terms, the question is whether we can find a family of observables that *separates the points of* S, in the sense that if two points s, s' of S are distinct, then there is already an observable f_i in the family that can tell them apart; $f_i(s) \neq f_i(s')$.

At this level of generality, the answer to the question depends only on the *cardinality* of S, i.e., on a property of S *as a whole,* and on the size of the family $\{f_\alpha\}$.

6C. Analysis and Cartesian Products

There are many ways to discuss Cartesian products. One of them involves only the ideas we have already developed in the preceding section, equivalence relations on a set. I will start with that.

Let us suppose that f, g are observables of S, i.e., that

$$f, g : S \rightarrow R$$

We have seen that we can *encode* every element of S uniquely into the pair of its values

$$s \rightarrow (f(s), g(s)).$$

Let us ask how we can *decode,* back from pairs of values to elements of S.

The question is: given a pair of *values* (numbers) $r \in f(S)$, $r' \in g(S)$ in the spectra of f and g respectively, does the pair (r, r') decode into anything? Is there an element s in S such that

$$f(s) = r, \; g(s) = r'?$$

Or put another way: do the R_f-class $f^{-1}(r)$ and the R_g-class $g^{-1}(r)$ have a nonempty intersection?

If the answer to this question is *no* for some pairs (r, r'), then such pairs cannot be decoded. If the answer is always *yes,* so that every R_f-class intersects every R_g-class and conversely, then every pair (r, r') of spectral values decodes back into S.

Now let us recall that

$$S/R_f \cong f(S)), \; S/R_g \cong g(S).$$

Let us ask how the quotient set S/R_{fg} is related to the spectra $f(S)$ and $g(S)$.

It is obvious from what I have already said that, if every R_f-class intersects every R_g-class and conversely, then S/R_{fg} *is* precisely the set of all pairs (r, r') of all spectral values of f and g respectively. That is, in this case, we can write

$$S/R_{fg} \cong f(S) \times g(S).$$

In other words: a Cartesian product is the same thing as a *pair* of equivalence relations that possess the property above. Likewise, when we have produced such a pair of equivalence relations, we can say that we have expressed or represented the set as a Cartesian produce of their quotients.

In fact, the discussion above says quite a bit more. Given any pair of equivalence relations on a set, we can always encode the elements of the set *as a subset* of the Cartesian product of the quotient sets. That is, we can always encode *into* this Cartesian product, but generally *not onto* the product. *Onto* requires the further condition I have stated above; only in this case can we unrestrictedly decode back from the product of the quotient sets.

What does the condition amount to, for equivalence relations R_f, R_g coming from observables? Its failure means that, if $f(s) = r$, the *value* $g(s)$ cannot be arbitrary; it is already at least partially determined by $f(s)$. Intuitively, the "information" about s residing in the value $g(s)$ is at least partially included in the "information" already conveyed by $f(s)$. I have called this situation (see *FM*) a *linkage* between f and g. Only if f and g are *unlinked,* so that what is general with respect to f is special with respect to g, and conversely, will this condition obtain. As we will see later, this general idea of linkage between observables is closely tied to the ideas of *constraint* and *recursion,* which we have already discussed (see section 4F). It pertains to the *ranges* of observables and to entailment in these ranges.

From this point of view, then, a set S *is* a Cartesian product if we can find a pair of unlinked observables f, g. We then have the projections II_f, II_g respectively, and hence the diagram

$$S/R_f \overset{\Pi_f}{\leftarrow} S \overset{\Pi_g}{\rightarrow} S/R_g.$$

Further, if U is any other set, and if

$$\varphi : U \rightarrow S/R_f,\ \psi : U \rightarrow S/R_g$$

are any mappings whatsoever, then there is a uniquely determined map

$$\theta : U \to S$$

which makes the diagram

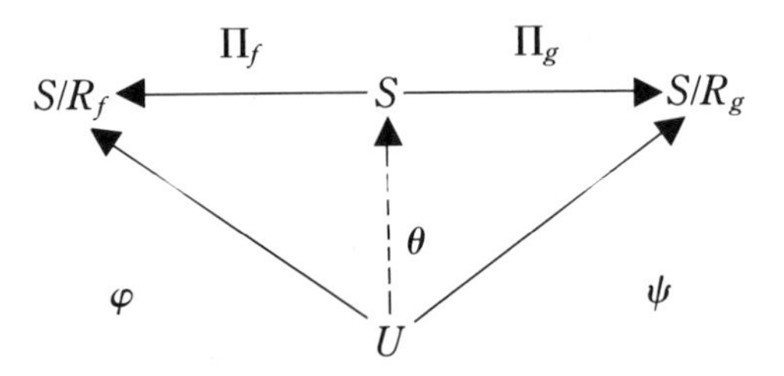

commute; put

$$\theta(u) = (\varphi(u), \psi(\theta)).$$

for every u in U.

This last is a *universal property,* which nowadays is often taken as the *definition* of Cartesian product. It intuitively expresses the Cartesian product as a kind of minimal thing, the smallest set that projects onto its factors (in the sense that any other set U that maps onto these factors can be made to go uniquely through the Cartesian product itself).

In what follows, I shall call any expression of a set S as a Cartesian product of quotient sets an *analysis* of S. The result of such an analysis is obviously to encode the points of S into a family of values, i.e., into a set of *coordinates,* which are the values of suitable observables of S. In turn, these coordinates are individually the values of certain invariants for equivalence relations on S.

Now let us see what happens when we compare two such analyses. We have already noted that the set $R(S)$ of all equivalence relations on S has a lot of structure. The intersection of any two equivalence relations in $R(S)$ is again an equivalence relation (the mutual refinement); under the natural inclusion relation the refinement is bigger (has more classes) than the relations it refines. We can export this refinement relation to analyses of S in the obvious fashion, so that *the analyses themselves can be partially ordered.*

When we compare two such analyses, we are in a situation we have seen before: namely figure 6C.1.

The problem is to specify the conditions under which we may find the dotted arrows in figure 6C.1 that make the entire diagram commute. That is: when can we encode one *analysis* into another, in such a way that one of them can itself be expressed as a model of the other?

The answer should now be very easy. *We can do this if, and only if, one*

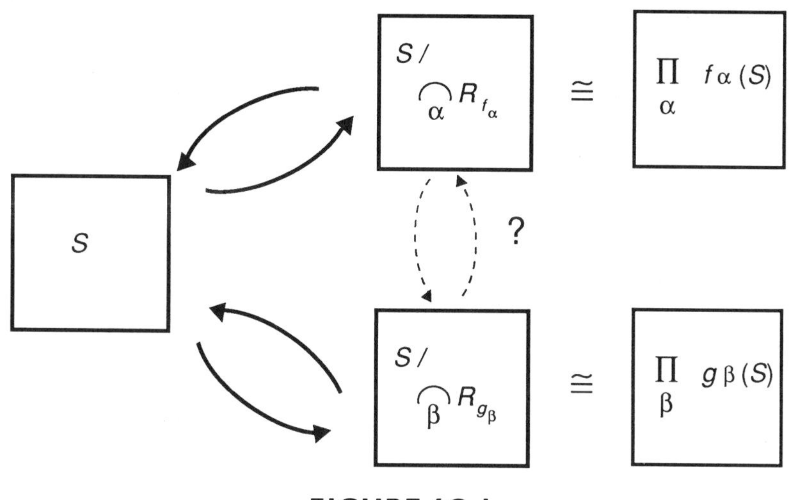

FIGURE 6C.1

of the analyses refines the other. In terms of the ordering relation on these analytic models, one of them (the refinement) has to be larger than the other. We can then encode by projecting down from the refinement (the larger one) to the smaller one; it is immediate to verify that, under these conditions, the entire diagram commutes. I leave the details to the reader.

On the other hand, if one of the analyses does not refine the other, so that neither is larger or smaller than the other, there is no way to encode one into the other in a way that preserves the modeling relation. Such analyses are thus *incomparable;* they may loosely be said to individually encode different *kinds* of information about the set S. There is accordingly no way to directly compare the analyses themselves, no way to formally construct one from the other. Put yet another way, there is no *entailment,* no *linkage,* between such incomparable analyses.

Given two such incomparable analyses, however, we can always construct from them a new, bigger one that refines them both, as we have seen above. Namely, we can form a new product, whose equivalence relation is the intersection of the given ones. Specifically, if we have the two analyses

$$S/\bigcap_{\alpha} R_{f_\alpha} = \prod_{\alpha} f_\alpha(S)$$

$$S/\bigcap_{\beta} R_{g_\beta} = \prod_{\beta} g_\beta(S)$$

we can form a new one, refining both of them, by putting

$$R = (\bigcap_{\alpha} R_{f_\alpha}) \cap (\bigcap_{\beta} R_{g_\beta})$$

and noting that

$$S/R = (\prod_{\alpha} f_\alpha(S)) \times (\prod_{\beta} g_\beta(S)).$$

There are some technical details to be addressed, arising from possible redundancies (linkages) between the observables f_α and g_β, but they are not important for our present purposes, and I omit them here.

Let us pause to sum up what we have learned so far. Our purpose, as we recall, is to learn something about the modeling enterprise per se, by looking at mathematical or formal instances of it. We have been looking in particular at mathematical metaphors for measurements (observations), which transmute into (real-valued) mappings, their values and their spectra. The important things for us were these spectra, which, as we have seen, are labels for equivalence classes of elements indistinguishable to such a mapping or observable. We have seen how every *family* of such observables gives rise to an analysis, a representation of the set on which the elements of S are characterized in terms of the Cartesian product of the spectra of those observables.

Every such analysis thus gives us a model, in particular, what we have called an *analytic model.* As we have seen, the totality of such analytic models can be identified with (1) the totality of equivalence relations on the set S, and (2) with the totality of *sets* of observables, the set of all subsets of $H(S, \mathbf{R})$. Thus, given S, the set of all its analytic models is a well-defined thing, another set that I shall henceforth denote by $\mathbf{A}(S)$.

As I have emphasized, this set $\mathbf{A}(S)$ of all analytic models of S has a lot of internal structure, which formally persists *even if we forget about* S. In particular, $\mathbf{A}(S)$ enjoys a partial order relation (refinement), and an intersection operation that allows us to build a bigger model in $\mathbf{A}(S)$ from two given ones. Furthermore, when one model in $A(S)$ refines another, we can establish a modeling relation *between them.*

I may sum all this up by saying that $\mathbf{A}(S)$ is a category. Its *objects* are precisely the analytic models of S, which may be concretized in terms of the equivalence relations on S, or in terms of sets of real-valued mappings on S. If M_1, M_2 are objects in this category, we say there is a *mapping* between them if and only if one of them refines the other. It is clear that composition of such maps is associative.

The thrust of our discussion has been to pass *from S to* its models.

Obviously, given any M in $\boldsymbol{A}(S)$, S realizes M. But S is not the *only* realization of M. I invite the reader to begin to ask questions like: what can we say about the *totality* of realizations of M? Or: given a *subset of* $\boldsymbol{A}(S)$, what about the totality of their simultaneous realizations (of which S itself is obviously one)? Most generally: in what sense, if any, can we say that S itself is *uniquely* determined by $\boldsymbol{A}(S)$, the totality of its analytic models? I will not pause to examine such questions in detail here, but they will become increasingly important as we proceed. It should be clear that the epistemological correlates of such questions are very deep indeed.

In any event, an analytic model of the type I have described should look very familiar to anyone versed in the science. It looks essentially like a *state space;* the observables that define its constituent quotient sets are *state variables.* By making such models, we pull questions about S itself into corresponding questions pertaining to the *ranges* of these state variables; we have done this kind of thing before, particularly in our discussion of *recursivity* (see section 4C above). Indeed, the proverbial astute reader should have already noticed that we are now using the same word ("state") that was originally introduced (see section 4H above) exclusively in connection with recursivity, even though in our present discussion there is as yet no recursivity explicitly in view. This apparent equivocation is neither accident nor carelessness (at least, not on our part); I will seek to remove it in due course.

Conversely, we *impute* properties of our analytic models back to the domains on which its observables or state variables are defined, by decoding formal properties of their images back to the domain. Price among these is the imputation of a notion of *distance* or *metric* back from such a model to its realizations, in effect by mandating that the natural projections of S onto its quotient spaces all be *continuous.* This in turn allows us to introduce the basic notion of *stability* (especially *structural stability*) into our considerations. I shall not pause to pursue these considerations here, but they are fundamental, and we will meet with them again as we proceed.

Note also that, at the level of generality we have maintained above (i.e., where S is just a set), we could allow an observable to be any real-valued mapping on S. If there is additional structure on S (e.g., if S is an algebra, or a geometry), then we need to respect that additional structure by limiting, or *constraining,* our choice of observables in some fashion. We may do this in several ways; the simplest, of course, is to throw away all observables that do not respect the additional structure (i.e., by passing to an appropriate subset of the set of all real-valued mappings on S). Alternatively, we may introduce *relations* between observables (e.g., by calling one of them the *derivative* of another). These are, however, relatively technical issues and do not affect the essential features of analytic models I have sketched above.

The essence of analytic models is that *every element s of S gets encoded* into something; in fact, it gets encoded into the values $\{f_\alpha(s)\}$ of the observables $\{f_\alpha\}$ we have chosen to define the model. We do not have a situation in which a *part* of S, a piece or subset of S, gets encoded, and the rest of S does not. Of course, we can in purely formal terms look at some subset U of the range $f(S)$ of an observable; then $f^{-1}(U)$ is, of course, a union of equivalence classes of R_f and is hence a subset of S encoded into U. But from the analytic point of view, there is no *intrinsic* significance to pieces or subsets of S isolated in this fashion. I now turn our attention to a quite different circle of ideas, which also lead us to models of S, but in which certain distinguished pieces or subsets play the central role from the beginning. The essence of this approach is to encode such pieces, and from these to construct an encoding of S itself.

6D. Direct Sums

In the preceding section, I introduced the concept of the *analysis* of S; the expression or encoding of a set S into a Cartesian or *direct product* of *quotient spaces*. I now turn to another circle of ideas, the expression of S as a *direct sum* of *subspaces*. As I noted earlier, these two circles of ideas are dual to each other, in a precise category-theoretic sense, but they are generally very different.

In the present section, I show that ideas of direct sum (or *coproduct,* as category theorists would prefer to have it) also lead to models. Since the idea here is to build a set S up by assembling pieces, rather than to break it down into coordinates, I call the resulting models *synthetic*. As we noted in section 6A above, we will find that many of the deepest difficulties in science find their root in the presumption of relations between analytic and synthetic in this sense.

Indeed, the essence of a synthetic model is precisely that it is built up out of disjoint, separate pieces. Hence it can be "unbuilt" the same way; it can in effect be cut apart again with scissors. Analytic models, on the other hand, are built from observables that, in effect, see the system whole. Thus, if f, g are any "analytic" observables, there is generally no way to separate $f^{-1}(r)$ and $g^{-1}(r')$; they generally overlap and cannot be regarded as direct summands in any sense.

Even in elementary Euclidean plane geometry, overlapping figures have always been regarded as troublesome. In Euclid's own *Elements,* for example, a proof involving overlapping triangles appears at the very beginning (book I, proposition 4). Because of the appearance of the associated figure, it was long called *pons asinorum,* the bridge of fools. Euclid obviously chose this proof deliberately, instead of the proof that everyone learns in high school geometry and that does not involve overlapping triangles. Why? The usual view is that then, as now, mathematics was regarded as an esoteric discipline; a "hard" proof at the beginning would discourage all but the adept. In my view, it is equally likely that Euclid chose the "hard" proof to show than one could separate, in thought, properties that could not be separated in the mundane, corrupt world of paper and scissors. Thus the *pons asinorum* does more than simply prove a theorem; it illustrates the possibility that what we have called "analytic" and "synthetic" need not coincide.

In the preceding section, the basic strategy was to learn about S by *projecting* it onto quotients, using the vehicle of observables. As we saw, we learned about S by looking at the ranges of mappings defined *on* S, and imputing back *from* range *to* domain. The strategy now will be: to learn about S by *injecting* things into it; to consider S itself as range rather than as domain. Accordingly, in this strategy, we seek to impute properties to S from what is injected into it, rather than what is projected out of it.

In epistemological terms, such a strategy does not look promising. In the natural world, S is not known; it is the object of study, to be probed through observation and modeling. As such, it seems naturally to be *domain,* on which observables are defined, and from which modeling relations are established and interrogated. To treat S as range seems to presuppose, from the outset, a knowledge of S that is most difficult to justify on any phenomenological grounds whatever.

On the other hand, we like to atomize. We like to build complicated situations out of the elementary or simple ones. We have a predilection for modularity, for taking a set of elementary pieces, putting them together in all possible ways, and *positing* that this exhausts reality. Indeed, formalization in mathematics is nothing but this. So, in its way, is algebraic topology; see section 5I above. If we are correct in this position, then the modeling of any aspect of reality consists of no more than rummaging through our constructs to find the one or ones that correspond to it, i.e., with which a modeling relation can be established. This in turn requires merely an identification of the "right" set of elementary pieces and the "right" ways of putting them together, i.e., the right *syntax.*

Seen in this light, it is clear that we can indeed learn much by pursuing this strategy. If it succeeds with respect to a system S, we can say that we learn something about S from its success. If it fails, we learn something else important about S; even more, we learn something about the strategy itself. This in fact is my own idea of a "good" problem, one from which one learns something no matter how it comes out.

Since this kind of strategy is therefore audacious in the extreme, there is all the more reason for contemplating it metaphorically, entirely within the formal realm. Indeed, a great deal of mathematics, including the seeds of Category Theory itself, has arisen entirely from the pursuit of such a strategy. This is the task of the present section.

I can start purely formally, with the universal definition of direct sum, or coproduct. Let us suppose that U, U' are (disjoint) sets. Then S is called their *direct sum*

$$S = U + U'$$

if we can produce mappings *(injections)*

$$\begin{aligned} i &: U \rightarrow S, \\ i' &: U' \rightarrow S \end{aligned}$$

such that, if $\varphi : U \rightarrow X$, $\psi : U' \rightarrow X$ are any mappings, there is a (unique) mapping $u : S \rightarrow X$, which makes the diagram

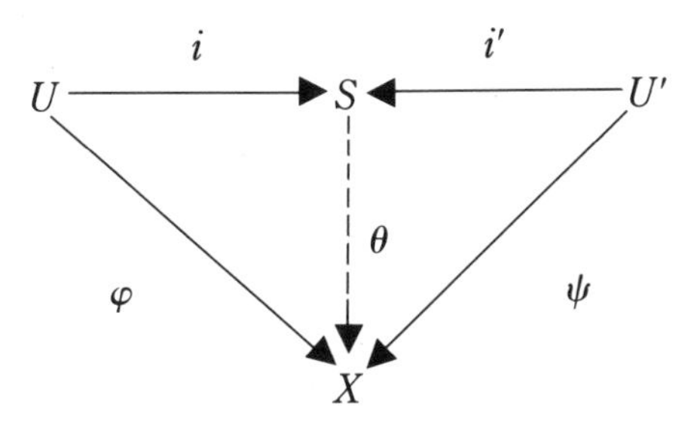

commute. (Intuitively, define u by

$$\begin{aligned} \theta(s) &= \varphi(s) \quad \text{if } s \in U; \\ &= \psi(s) \quad \text{if } s \in U') \end{aligned}$$

This universal property identifies the direct sum as, in a sense, the smallest set containing both U and U'. It is, of course, reminiscent of the universal diagram for defining direct or Cartesian products, except that all the arrows (mappings) are turned around; instead of projections of S onto its factors, we have injections (intuitively, inclusions) of summands (subsets) into S. This "turning around" of arrows in a diagram is a precise manifestation of duality in Category Theory; concepts come in dual pairs of this kind, and the dual of such a concept is denoted with the prefix "co-" (hence "coproduct" for direct sum, the dual of "product").

We can clearly produce the direct sum

$$S = \sum_{\alpha} U_{\alpha}$$

of *any* family of disjoint sets U_{α}, using this same universal definition.

The disjointness of the $\{U_{\alpha}\}$ means that their direct sum S is automatically partitioned into nonintersecting subsets, which can be thought of as the equivalence classes of some equivalence relation R_{σ} on S. The quotient set S/R_{σ} is in 1–1 correspondence with the index set $\{\alpha\}_i$; the general property embodied by R_{σ} is that of belonging to the *same* subset U_{α}. In these terms, to identify a point of S means (1) to identify the subset U_{α} to which it belongs (i.e., specify its equivalence class under R_{σ}), and then (2) locate it specifically within that class. Seen in this light, any direct sum comes with a distinguished equivalence relation on it; conversely, any equivalence relation on a set allows us to express that set as the direct sum of its equivalence classes.

We will now think of each U_{α} as a system in its own right and use the ideas just developed to construct a new system S. The relation of S to each of the U_{α} from which it is built is analogous to the relation between molecule and atom or between population and individual. Of course, this is not an accidental or incidental observation; it is in fact the essence of the direct sum and of the synthetic approach itself.

Thus, let us take a family of disjoint sets U_{α}. We explicitly allow some of these sets to be *copies* of others, identical (congruent) in all features, but consisting of different points. Indeed, we shall ultimately suppose that we have at our disposal as many copies, in this sense, of any U_{α} as we need or desire. The only limitation is that, *as sets,* these copies are disjoint.

Since each U_{α} is a system, we shall suppose that U_{α} can be represented in the form

$$M(U_{\alpha}) = \prod_{i} U_{\alpha}/R_{f_i\alpha}.$$

That is, each U_{α} is assigned a fixed *analytic* model in terms of observables $f_{i\alpha}$ defined *on it.* Of course, we assume that if U_{α} and $U_{\alpha'}$ are copies of each other, then so are the assigned analytic models $M(U_{\alpha})$, $M(U_{\alpha'})$. Intuitively, then, each point of the model $M(U_{\alpha})$ represents or encodes a *state* of U_{α}, with respect to the *state variables* $f_{a\alpha}$.

Let us note explicitly that, if $\alpha \neq \alpha'$, then any two observables $f_{i\alpha'}$ $f_{j\alpha}$, belonging to U_α, U_α' respectively, are *automatically unlinked.* They are defined on different sets, and at this level there is accordingly no kind of relation that can exist between the points of their spectra. This absence of linkage obtains, even if U_α and $U_{\alpha'}$ are copies of each other, and even if $f_{i\alpha}'$ $f_{j\alpha'}$ are "the same observables" in $M(U_\alpha)$, $M(U_{\alpha'})$ respectively.

We will now assign a model $M(S)$ to the direct union or direct sum

$$S = \sum_\alpha U_\alpha.$$

Intuitively, a *state* of S will be determined by assigning a state to each summand U_α. That is, $M(S)$ consists of all α-tuples (u_α), where $u_\alpha \in M(U_\alpha)$. Accordingly, we shall also write

$$M(S) = \sum_\alpha M(U_\alpha).$$

Since this is obviously a very important kind of construction, we will pause to visualize it in various ways.

First, let us recall that a direct sum is automatically partitioned into equivalence classes of an equivalence relation R_σ. Each class of R_σ is precisely a summand of the direct sum. As always, we can think of the quotient set S/R_σ as a base space, and the equivalence classes themselves (in this case, the U_α) as *fibers.* We thus have a picture of the type exhibited in figure 6B.1 above. In the present case, we can replace these original fibers U_α by their *models* $M(U_\alpha)$. Thus we keep the original base space S/R_σ, but the individual fibers now consist of the *states* of U_α, the images or encodings of U_α in $M(U_\alpha)$.

As we noted, a *state* of the direct sum S is to be imaged or encoded by choosing one state from each of the direct summands U_α of S. If we do this, we have thereby constructed a *cross-section* of the fibers, as indicated diagrammatically in figure 6D.1. Each such cross-section is allowed, because the models $M(U_\alpha)$ are all unlinked. $M(S)$ accordingly consists of all such cross-sections.

I pause here to draw attention to one crucial feature of this whole circle of ideas. Namely, if S is any system constructed in this fashion, *then the encoding of S into $M(S)$, which I have described above, is already entailed by the construction itself.* I have stressed repeatedly that encoding of natural system into formalism is *generally unentailed.* The whole purpose of direct sums, in a sense, is precisely *to create a situation in which encoding is itself entailed.* That is why people like direct sums and why they have acquired such a dominant place in theoretical science. For in this situation, and in it alone, we may replace *art* by *craft.* And it is always much easier to be a craftsman than an artist.

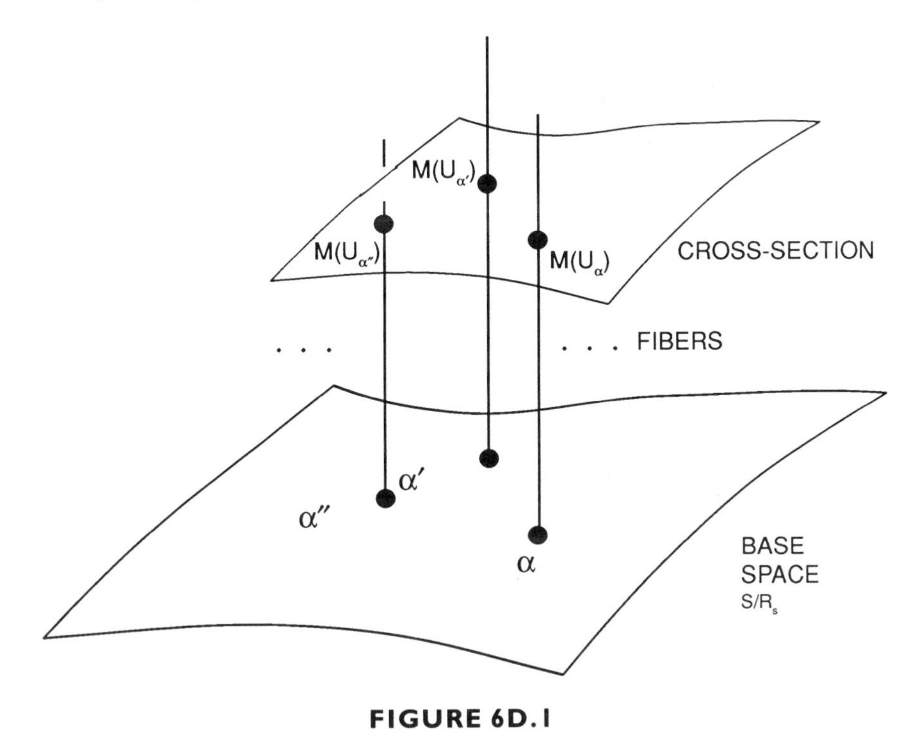

FIGURE 6D.1

We may also note here that the causal status of these synthetic states is very different from what we have seen before. Being a construct, a synthetic state $s \in S$ is itself an effect, something entailed; it is not the answer to a question but something we ask questions about. For instance, we can now ask "why *s*?" a question we cannot even ask in terms of analytic models but that is now meaningful. One kind of answer to this question is: "because u_α". In this way, causal significance devolves upon the direct summands; more specifically, their elements are *material causes* for the cross-sections they generate.

It is clear that, given any family U_α of disjoint subsets, we can build a lot of systems S out of them using the ideas above. For we can choose (1) which ones among the given family we will actually use as summands and (2) how many *copies* of each we will use (*"ploidy"*). In fact, as we shall now see, we can construct a whole *category* of them. Moreover, the *models* of these systems come equipped with a *natural ordering,* which embodies one of the many notions of *subsystem.*

To begin with, let us conjure a bit with this idea of "ploidy." Let us suppose for simplicity that we begin with a family U_α whose elements are *all different* (i.e., no two are copies of one another), and let us denote the *index set* for this family by A. As usual, we know that

$$A = \sum_{\alpha \in A} U_\alpha / R_\sigma.$$

If we want to use *copies* of these U_α, we will need bigger index sets than A, because we will also need to index or count the copies. For instance, if we want to use two copies of the same summand U_{α_o}, we will need a new index for the second copy. Why not call it 2_{α_o}? If we use three copies of U_{α_o}, we will need another index for the third copy; why not call it $3\alpha_o$? It is then obvious that *both* (1) the selection of particular summands from among the distinct sets U_α and (2) how many copies of each we want to use can be associated formally with expressions like

$$n_1\alpha_1 = n_2\alpha_2 = \ldots$$

where the n_i are integers, and the α_i are elements of A. It is convenient to allow the integers n_i to be either *positive or negative,* where we can think of a negative coefficient as corresponding to the *deletion* of copies of a particular U_α from a family.

The totality of expressions of this form, which intuitively represents the totality of ways of choosing any number of copies of constituent direct summands from among our given family $\{U_\alpha, \alpha \in A\}$, itself comprises a familiar mathematical object; it is the *free abelian group generated by A,* or the *free* ***Z****-module generated by A*. Let us denote it by $\mathbf{G}(A)$. There is obviously a 1-1 correspondence between the elements in $\mathbf{G}(A)$ and the totality of systems S we can build from the sets U_α, where A indexes the U_α.

Now $\mathbf{G}(A)$ automatically comes equipped with a natural ordering relation, a partial order, derived from the ordering relation on the integers. Namely, if

$$g_1 = \sum_{\alpha_i \in A} m_i d_i,$$

$$g_2 = \sum_{\alpha_i \in A} n_i d_j$$

are elements of $\mathbf{G}(A)$, we will say

$$g_1 \leq g_2$$

if and only if

$$m_i \leq n_i$$

for each $\alpha_i \in A$. That is, $g_1 \leq g_2$ means that the associated systems S_1, S_2 are built from the same summands U_α, but the *ploidies* m_i of these summands in S_1 are smaller than the corresponding ploidies n_i in S_2.

It is immediate now to verify that, if $g_1 \leq g_2$ in $\boldsymbol{G}(A)$, and if S_1, S_2 denote the respective systems they represent, then there is a natural inclusion of $M(S_1)$ into $M(S_2)$; an injection

$$\eta : M(S_1) \rightarrow M(S_2)$$

which expresses $M(S_1)$ itself as a direct summand of $M(S_2)$. The details are elementary and a good exercise in the ideas I have developed so far; I leave them to the reader.

From this, it is further immediate that the totality of the $M(S)$ that can be built from our given family $\{U_\alpha\}$, $\alpha \in A$, form a category. The objects in this category are the $M(S)$ themselves, and the mappings are just the injections η we have just defined.

In fact, this category has a lot of structure. One can see a lot of linearity floating around in it, arising directly from the properties of direct sums, and the close relation of our category to an abelian group or $\boldsymbol{Z}$-module. We could spend a great deal of time in merely investigating this structure, and its epistemological correlates, which have themselves become elevated to the status of Laws of Nature. I shall not do so explicitly here, because as we shall see, they are far from that. But the reader is invited to play with these ideas at leisure, if so inclined.

6E. Analytic and Synthetic: Comparison and Contrast

In the preceding sections I have developed two entirely different, dual ways of looking at systems (or at least, at two formal analogues of the modeling process for systems). Now we are going to put these two together, and see what happens when *analytic and synthetic models of the same system are presumed to be available.*

The basic question is one we have seen before. We can put it in the form of a diagram (figure 6E.1). Here, S simultaneously realizes two different *kinds* of model. The question then is: under what circumstances can we establish relations between the models, as symbolized in the diagram by the dotted arrows? In formal terms, these arrows amount to establishing a relation between direct products and direct sums. As it turns out, this is perhaps the central question in all of theoretical science, for, as

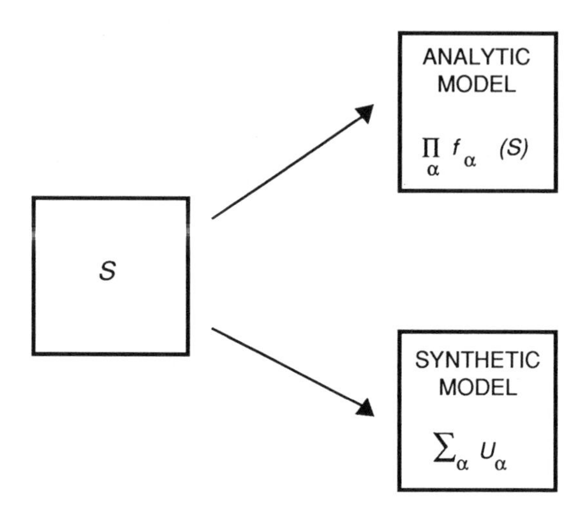

FIGURE 6E.1

we have seen, synthetic models are an embodiment of pure syntactics, whereas analytic models are inherently semantic.

The key to establishing such relations between analytic and synthetic models of S lies again in the concept of *observable.* Observables are, of course, at the very core of analytic models, via the encodings

$$s \rightarrow (f_{\alpha}(s)) \qquad [6E.1]$$

of elements of S into sets of observable *values.* In synthetic models, on the other hand, we find an intermediate construct, namely, an expression of s in a new form

$$s = u_1 + u_2 + \ldots + u_r, \qquad [6E.2]$$

where $u_i \in U_i$. It is this construct that is interposed between what is encoded (namely $s \in S$) and the observable values that actually constitute the encoding. We shall now see what this means; specifically, we shall see that the imposition of a direct sum structure on S mandates a similar structure on the observables themselves. Intuitively, it requires that an arbitrary *observable* of S be generated by, or built up from, *observables that look only at summands of* S.

The reader may have noted that, so far, we have not invoked any of the arithmetic structure that $H(S, \boldsymbol{R})$ inherits from the real numbers $\boldsymbol{R}$. At the very least, $H(S, \boldsymbol{R})$ is a *ring;* we can add observables by adding their *values:*

$$(f_1 + f_2)(s) = f_1(s) + f_2(s)$$

likewise, we can multiply them, multiply them by real numbers, etc. We have heretofore not mentioned these structures for a very good reason: namely, in analytic encodings, they have *no significance;* they do not *decode* into anything about S. We can see this if, for instance, we ask how the spectrum $(f_1 + f_2)(S)$ of the sum of two observables is related to the spectra $f_1(S)$, $f_2(S)$ of the summands. Or, what is the same thing, we ask how the equivalence relation $R_{f_1 + f_2}$ is related to the individual equivalence relations R_{f_1} and R_{f_2}. The reader can easily be convinced that in fact there *are* no relations in general, and hence the absence of any significance of algebraic operations on the values of observables for what these observables encode. See, e.g., *FS* for a more nearly full discussion.

For direct sums, however, the situation is completely different. In a nutshell, this is because observables defined on different summands of a direct sum have different *domains;* adding two such creates in effect a new observable, *defined on a bigger domain* (namely, on the direct sum). We can thus build observables on S itself from observables defined only on its summands, and this construction *is* reflected in the algebraic structure of $H(S, \boldsymbol{R})$ in this case.

For instance, if we have

$$S = U_1 + U_2,$$

then an observable f of U_1 alone *is* an observable of S as well; an observable g of U_2 likewise *is* an observable of S. If we conventionally write

$$F = (f, 0),$$
$$G = (0, g),$$

then F and G become observables of $S;$ in effect, the observables f, g, defined on U_1, U_2 respectively, are "lifted" from their initially different domains to a new common domain (namely S); by adding them, according to

$$F + G = (f, 0) + (0, g) = (f, g)$$

we thereby produce a new observable on S.

In this way, given an expression of S as a direct sum

$$S = \sum_{\alpha} U_{\alpha}$$

we can attach meanings to expressions of the form

$$f = \sum_{\alpha} r_{\alpha} \varphi_{\alpha}$$

where the r_{α} are numbers, and the φ_{α} are observables of the individual summands U_{α}. Each such f is in fact an observable of S; its values are determined for any s in S by the values assumed by the φ_{α} on S. But by hypothesis,

$$s = \Sigma u_{\alpha},$$

we have explicitly

$$f(s) = \sum_{\alpha} r_{\alpha} \varphi_{\alpha}(s) \equiv \sum_{\alpha} r_{\alpha} \varphi_{\alpha}(u_{\alpha}). \qquad [6E.3]$$

From this, it is easy to see why linearity plays such a central role in synthetic modeling.

On this basis, we can now formulate our central conclusions regarding the relations between synthetic and analytic models. Our first is:

Every synthetic model is an analytic model.

In fact, a synthetic model is of a very special type, in terms of the observables that define it. Specifically, these observables must either look only at individual summands or be expressible as a linear combination of such. That is: the values of observables defining a synthetic model are determined, or entailed, entirely by values on summands.

But:

There generally exist analytic models that are not synthetic models.

This is in fact the main result we have been driving at and that we claim is perhaps the central one in the epistemology of theoretical science. As befits this status, it is a little subtle and involves a number of aspects, which we should separate.

First, we should observe that, *given* an expression of S as a direct sum of summands U_{α}, there are always plenty of observables in $H(S, \boldsymbol{R})$ that are not of the form [6E.3], i.e., cannot be expressed as linear combinations of observables φ_{α}, which look at the summands U_{α} alone. Such an observable f can always be chosen in such a way to split the φ_{α}-equivalence classes; i.e., so that I can find elements s, s' in S for which

$$f(s) \neq f(s')$$

although

$$\varphi_\alpha(s) = \varphi_\alpha(s')$$

for every α. Since we can always use such observables to encode S analytically, it follows that there are always analytic models of S that have no synthetic counterpart in terms of *those summands* U_α.

On the other hand, it remains possible that, *given* an analytic model of S, we can *find* summands V_β such that

$$S = \sum_\beta V_\beta$$

and such that every observable in the analytic model can be expressed as a linear combination of the form [6E.3]. This is, of course, quite a different matter. We are at the moment claiming only that, in general, there is no direct sum decomposition of S that turns *every* analytic model of S into a synthetic model. In more precise terms: *there exist analytic models that possess no synthetic refinements.*

We see in any event that, in general, *generically,* synthetic and analytic models do not coincide. We recall, however, that we are still in a formal world, which is at our disposal in a way that the external world is not. In particular, we can *require* of our formal world that synthetic and analytic *do* coincide. This is, of course, *a very special formal world;* it is *defined* by a property that, as we have just seen, is not satisfied in general. It is therefore *degenerate.* And we can ask what this very special world is like.

In brief, it is a world in which S *itself is already linear in some sense;* indeed, a very strong sense. It is tantamount to mandating that S itself is, in effect, an *inherently syntactic structure.* In such a world, every world can be expressed as a direct sum of fixed "atoms" or summands. We will see later the dire consequences of assuming, at the outset, that synthetic and analytic models coincide (see chapter 9). We merely remark, at this point, that presuming anything like this about S is tantamount to placing the most severe restrictions on causality itself, in effect, forcing causal relations to themselves be constrained by something that is itself uncaused. This in turn far transcends the entitlements we obtain from Natural Law, and hence, requires the most careful scrutiny.

6F. The Category of all Models of *S*

In our preceding discussion, we have seen the basic distinction between what I have called analytic and synthetic modeling. One analytic model is called better than another if it is a refinement, if it discriminates better between the elements of S. On the other hand, a synthetic model is called better than another if it "sees" more summands. This idea of "better" was reflected formally in the ordering relations we found on the totality of analytic and synthetic models of S.

As our preceding discussion shows, these two ideas of resolution do not in general coincide. Nevertheless, we should be able to put them together into a single coherent framework, which reflects the fact that both approaches lead to encodings (models) of S. We can also see that this larger framework should also encompass "hybrids" between purely synthetic and purely analytic models, situations in which we can increase discrimination (i.e., generate a bigger [better] model) both by adding more observables (increasing our resolution) *and* by seeing more summands. These ideas lead us to what we shall call $\boldsymbol{C}(S)$, the category of all models of S.

The discussion that follows is designed to be generally valid, independent of any presumed relations between direct products and direct sums. The basic intent is to show that the characteristics of $\boldsymbol{A}(S)$, the category of all analytic models of S, extend in a natural way to $\boldsymbol{C}(S)$, the category of all models of S. These include (1) a natural partial ordering, which allows us to say when one model in $\boldsymbol{C}(S)$ is bigger (better) than another, and (2) an operation that allows us to put two models of S together to get a bigger one.

As I have repeatedly emphasized, the essential feature of a model of S is that it encodes something about S. In an analytic model, we encode an element s of S into a set of observable values

$$s \to \{f_\alpha(s)\}$$

and thereby encode S itself into a Cartesian product

$$s \to \prod_\alpha f_\alpha(S) = \prod_\alpha S/\bigcap_\alpha R_{f_\alpha}.$$

In a synthetic model, on the other hand, an element s of S is encoded into a cross-section of a direct sum:

$$s \to \{u_\beta\} \in U_\beta$$

and hence

$$S \to \sum_\beta U_\beta.$$

The essential fact about such encodings, which allows us to compare them even though they are of fundamentally different characters, is that they are in general *many-to-one;* that is, many elements of S generally get encoded into the *same* name or label in the model. Each such encoding, regardless of its specific character, thus defines *an equivalence relation on S;* two elements of S are equivalent with respect to a model if the associated encoding cannot resolve them, and hence, assigns them the same image or label in the model.

Thus, if $M(S)$ is a model of S, and

$$\varphi_M : S \to M(S)$$

is the associated encoding, there is an equivalence relation $R_{\varphi(M)}$ on S that is imposed by the encoding.

Thus, if M_1, M_2 are two different models of S, of whatever nature, we can compare them by comparing the equivalence relations they impose on S; in particular, we can partially order the equivalence relations on S. Indeed, from this perspective, we can see a very close relation between $C(S)$, the category of all models of S, and the set of all equivalence relations on S.

In the situation where S is a *natural* system, in which we are attempting to learn about S from its models, none of these ideas are available, for in this situation, we have no prior knowledge about what S is, against which we can compare what different models are telling us about S. In particular, we must decide whether one model is bigger than another on the basis of the properties of the models themselves, something that is generally not easy to do. Nevertheless, one of the values of the formal metaphor for modeling, which we have been examining, is that it illuminates these questions, by putting S and $M(S)$ on the same footing. We are thus enabled, at least metaphorically, to penetrate the black box S and see what a specific model $M(S)$ actually tells us about S.

Having established the natural partial ordering in $C(S)$, we can now turn to the question of putting two models in $C(S)$ together to get a bigger one. From our metaphorical point of view, this means simply generating from them a new model, imposing on S an equivalence relation that is the intersection (the mutual refinement) of the equivalences on S coming from the given models.

In the case of analytic models, we have already done this; given any two analytic models, with their respective encodings, we know how to encode S into a bigger model that mutually refines the given ones. Let me briefly describe the analogous situation for synthetic models.

Suppose then we are given a pair of synthetic models M_1, M_2 of S, neither of which refines the other. Each of them can be visualized as indicated in figure 6D.1, in terms of a base space (the summands) and the fibers over these summands. We can put them together by taking the union of the base spaces and keeping the fibers as they were before. Each element of S now encodes into a cross-section of the fibers over this larger base space. It is clear (*modulo* a few technical details, which we omit) that we obtain in this way a new synthetic model of S, which is in fact the mutual refinement of the ones we were initially given.

In general, we do not know whether S has any synthetic models at all. To show that there is a synthetic model, we must (1) construct such a formal structure and then (2) encode S into it in such a way as to establish a modeling relation. If there are synthetic models, it then becomes an item of great interest to see how they are distributed in $C(S)$, relative to the ordering relation in this category. For instance, we can ask: given an arbitrary model in $C(S)$, does it have a synthetic refinement? Questions of this kind, and their epistemological correlates, will play a central role in the considerations to follow.

I remind the reader that, when S is a natural system, Natural Law asserts that all we can know about S is embodied in its models, i.e., in what we have called $\mathbf{C}(S)$. Thus, we must build our knowledge of S from the formal structure of $\mathbf{C}(S)$. Thus $\mathbf{C}(S)$ itself becomes a kind of metaphor for S, for each individual S. And moreover, the totality of all these $\mathbf{C}(S)$, must be a kind of metaphor for the whole external world itself.

With these remarks, I will suspend for the moment our investigation into these formal metaphors for the modeling process itself. In this formal setting, we are able to see most clearly the distinctions between analytic and synthetic modeling. But note that we have not yet mentioned the word "relational," in the context of that general discussion.

In the next few chapters, we shall return to the world of natural systems, armed with the ideas we have developed. First, we shall see what it means to mandate, at the outset, that analytic models and synthetic models in some sense coincide, i.e., that the subcategory of synthetic models generates the category of all models. As we shall see, this takes us directly into the world of a certain class of natural systems, whose members we shall call *mechanisms*. Some of these mechanisms, those whose categories of models satisfy an additional restriction, will be called *machines*. It will turn out that these machines, as natural systems, are "organized enough" to admit relational models as well. But the limitations of mechanism (analytic = synthetic) turn out to place corresponding limitations on the kinds of relational models that machines can have, limitations that may be expressed in several ways but primarily in terms of causal entailments in such systems. When we lift those restrictions, which we may do in a relational contex+, we leave not only the class of machines but also the class of mechanisms. But we enter a world in which biology is possible, the world of life itself.

In order to do this, we still need one more circle of ideas, connected with the concept of *simulation*. It is to the development of these ideas that we shall now turn.

Chapter 7

On Simulation

THROUGHOUT THE preceding sections, I have often referred to the concept of the machine; particularly with respect to the machine metaphor of Descartes and the role it has subsequently played in biology. In the present section, I shall begin to sharpen the vague concept of machine, both as a formal concept and as a material one, within the same confines of Natural Law that we have used heretofore for everything else.

7A. The Machine Concept

It must be admitted, what was already obvious to Descartes, that some machines and some organisms do seem to share common attributes. As we have already noted, Descartes could perceive these commonalities, without any general conception of what a "machine" was, and even less conception of what an "organism" was, simply on the basis of casual acquaintance with a very few examples of each.

What were those general features that Descartes could perceive, on the basis of such limited and casual acquaintance, and that provided the basis for his claim that the one subsumed the other? Two things, I think: (1) machines and organisms, as Descartes knew them, exhibited the appearance, at least, of autonomous action, and (2) machines and organisms are both *organized,* heterogeneous, differentiated, composite entities. As to the former, the appearance of autonomous movement, without any necessity for an external, impressed force or push, was since ancient times a *diagnostic* for life, a touchstone to separate the animate from the inanimate. In those days, life was a *primitive,* in terms of which other phenomena of Nature found their explanations, and hence, the *animisms* so despised nowadays.

Indeed, it would never occur to an ancient that life was something that *needed* to be explained. As noted at the outset, it was in fact the rise of Newtonian mechanism, and its success in celestial mechanics, that provided the credibility for an entirely different view of the world, the "modern" view. In that new view, there was no room for a distinction between animate and inanimate; indeed, the distinction itself disappeared. It was only then that a need for an explanation of life became manifest; indeed, life was now to be explained in terms of the same mechanics that had previously explained the motions of the comets, the planets, and the stars, for there now *was* no other accepted mode of explanation.

Hence the allure of the machine metaphor. For if we could understand machines mechanistically, then we could understand organisms in precisely the same way. If nothing else, there is a most satisfying *parsimony* in this picture; biology in principle mandates nothing *new*. Hence no one has ever looked for anything new.

But what, exactly, is a machine? What is the intrinsic, distinguishing characteristic that separates machines from material systems that are not machines? The most obvious distinction, e.g., between a machine and a stone, is that the former is *fabricated* and the latter is not, but this is worse than useless for the machine metaphor. For one thing, it involves the history of a thing, rather than any intrinsic characteristic; for another, it invokes aspects of intentionality that are simply unacceptable in this context.

To avoid these difficulties, and yet retain the machine metaphor, there is nothing to fall back on except the *organization* to which I have already referred, something that Descartes could already notice. What distinguishes a material system as a *machine,* as distinct from a stone or a crystal, must somehow reflect its intrinsic organization. *Where* the organization has come from is now irrelevant; fabrication, and history in general, do not enter into the matter at all. Such concepts pertain to *a different order of question;* an order that we must not invoke in characterizing machineness per se, but that we must not forget, either.

But once we talk about *organization,* we are in a relational context. We are basically defining a machine as a material system that admits (i.e., that *realizes*) a relational description, a description in terms of sets and mappings, and entailments between them. In the present section, we shall explore the concept of *machine* from this point of view. In the process, we will get a better perspective on the machine metaphor itself, on how and why it ultimately fails. From this perspective, it will be easier to see what new is needed to cope with the phenomena of biology; indeed, some of the basic tools are now already in our hands.

7B. Some General Heuristic Remarks

The machines that so impressed Descartes were basically toys, constructed for amusement. They were a species of clockwork, though powered hydraulically rather than with a spring. When working properly, they exhibited a fixed routine or repertoire, which could be repeated endlessly, without change. It is this invariability that dominates the semantics of the term "machine" in common parlance, where it has a vaguely pejorative connotation. Ironically enough, the term (especially in the adjectival form) connotes lifelessness.

Following the Newtonian revolution, it was customary to refer to the heavens, and to the universe itself, as constituting one vast "clockwork," precisely because of the invariability and predictability with which Newton endowed them. These, as we saw earlier (see chapter 4), are of themselves consequences of the Newtonian encodings, whose principal feature is the entailment of what happens next by what is happening now, *recursiveness.*

It is indeed true that clockworks, as material devices, can be shown to share precisely this property of recursiveness. But there is a subtlety here that must be pointed out, for it will become very important later (see especially 7D below). Namely, in a clockwork, recursiveness is not manifested in the motion of the visible figures being manipulated by it; it arises deeper down, in the springs and gears that actually do the manipulation. If we change the gears, we change the repertoire of the figures. In other words, what the figures themselves are doing now does not at all entail what *they* will do next; rather, it is what the *gears* are doing now that entails what the *gears* will do next, and what the gears are doing entails what the figures are doing. This apparently trivial remark has in fact a number of significant correlates for system theory and system analysis, prime among them the germ of the concept of *program.*

It took a very long time for anyone to recognize that recursiveness, and program, were *formal concepts* that could be studied entirely in the abstract, divorced from any particular material embodiment. It did not, in fact, really occur until 1936, when Turing published his famous paper. In this paper, the term "machine" was for the first time systematically employed in an entirely formal context.

A most important, but rather subtle, issue arises precisely at this point, which will be very significant for us. Namely, the abstraction of concepts like program, and recursiveness, from the behavior of specific material systems, and the endowing of them with formal correlates, is itself in fact a particular instance of *modeling* (in fact, of *relational* modeling). Specifically, that means the establishment of a commutative modeling diagram (see figure 3H.2 above) in which the formalism sits on the right-hand side of the diagram; the natural or material system sits on the left. In the present case, it happens that the *formalism* has come to be called by the same name (*machine*) as that which it models. That name is applied, in fact, on the basis of *intrinsic characteristics* of the formalism, independent of any specific encodings or decodings that relate the formalism to any natural system. Thus, an equivocation is created; the unqualified term "machine" can now refer to a certain kind of material system, *or* to a certain kind of formalism, *or* to the relationships of modeling that might exist between them. This inherent equivocation in the term "machine" must be kept in mind as we proceed. By the time we are done, however, all of these diverse usages, and the relations between them, should be clearer.

The Turing machines are in fact the formal counterparts of the clockwork, although they were not constructed as such. In these "mathematical machines," the symbols play the role of the clockwork figures; the rules for manipulating symbols are the gears and springs, and the particular arrangement and disposition of these rules constitutes *program* and *algorithm.* On the formal side, we have seen all this before, the manipulation of symbols themselves without meaning or entailment, by means of arbitrary syntactic rules applied in arbitrary sequences; this is the essence of *formalization.*

In what follows, I shall take the point of view that a clockwork, whether natural (i.e., a machine) or formal (a formalization) is a *simulator.* A great deal of the present section is in fact devoted to clarifying what that means. *What* is being simulated depends on program, on the specific disposition of syntactic manipulation rules. What *can* be simulated is determined by (as we shall see, is limited by) the nature of the rules themselves and by the ways they can be strung together.

The relation we shall establish between simulator and that which it simulates is very different from any we have seen heretofore. In particular, it is very different from a modeling relation, with which it is customarily confounded. Simulation will turn out to play no role whatsoever in the formulation of Natural Law that we have developed; assertions to the contrary, which continue to be made from time to time, rest only on simple equivocations.

In a nutshell, the present section is devoted to establishing a characterization of *machine* in terms of the notion of *simulation.* Simulation is what machines do, and a system (formal or material) *is* a machine if it simulates, or can simulate, something else. I shall express these notions in terms of true modeling relations and shall be led thereby to a new class of systems, which I shall call *simple systems* or *mechanisms.* These are characterized by the property that *every model of them is simulable.*

I am thus using the concept of a mathematical machine and, in particular, its capacity to simulate, to specify a class of *natural* systems, namely, those whose models are all simulable. In this class, our ideas will suddenly slap together because, as I shall show,

EVERY MODEL SIMULABLE
IMPLIES
ANALYTIC = SYNTHETIC

and, to a sufficiently large extent, *conversely.* This central result will not only allow us to see, in the clearest possibly fashion, the intrinsic limitations of both machines and mechanisms as a basis for biology but also will tell us what to do about them. It will in fact lead us to a whole new epistemology, one that does not start from the presupposition, which characterizes contemporary physics, that every material system is a mechanism.

Subsequently, however, I will show that *there exist modes of organization whose material realizations cannot be simple.* Biology is full of them. In the process, we shall see explicitly how Rashevsky's Principle, his search for a relational biology, thereby leads directly to *new physics.* And we shall see much else besides, which pertains specifically to our basic question, "What is life?"

7C. The Algorithm

The essence of the concept of the "mathematical machine" is tied up with the notion of algorithm. In a certain sense, just as words can be thought of as molecules built up out of atomic letters, so algorithms can be thought of as molecules built up out of atomic syntactic operations on letters. There are a number of ways to develop the concept of algorithm; I will use an unusual one, exploiting the almost uncanny parallels between Newtonian particle mechanics (section 4H above) and formalization. As will be seen, at root these parallels arise because of their ties to recursiveness, to making what is to happen next be determined by what is happening now.

Let us first ask what can in fact happen in a formalized world. In such a world, we begin from a finite set of meaningless symbols, an alphabet

$$A_1 = \{a_2, \ldots, a_n\}.$$

These should be thought of as analogous to different *kinds* of particles, not as the particles themselves. The particles are identical *specimens,* available in unlimited numbers, each specimen belonging to its particular species in the alphabet. Invoking the customary abuse of language, I will refer to both a species a_i in A, and any individual specimen belonging to that species, as an alphabet symbol. What we are interested in is the characterization of certain populations or spatial arrays of alphabet symbols, namely, what we earlier called words.

It will simplify the discussion in some ways if we suppose that one of the letters of our alphabet A is the *empty letter e,* sometimes called a blank, or blank space. In Newtonian terms, this would correspond to the absence of a particle, this absence itself being treated as if it were itself a particle (the "empty particle").

In particle mechanics, we define a *configuration* by assigning to each individual particle in a family a specific location, or position, or coordinate. In the process, we define thereby a particular spatial pattern of the particles in the family. In that situation, our interest is in the particles, and hence, we naturally regard the location of a particle, its coordinate, as *an attribute of the particle.* However, if our interest is primarily in the pattern, it is more natural (and completely equivalent) to regard the particle that happens to reside at a particular coordinate as *an attribute of the coordinate.* Intuitively this means that, instead of riding with a moving particle as its coordinate changes, we can just as well sit at a coordinate and watch different particles move in and out. For our present purposes, it is convenient to adopt the latter view.

Let us now consider a word

$$w = a_{i_1} \ldots a_{i_r},$$

a concatenation of alphabet symbols, as a spatial pattern. If $I_r = \{1, 2, \ldots, r\}$ is the set of the first r integers, then we may think of w as a mapping (also denoted by w) of the form

$$w : I_r \rightarrow A$$

defined by

$$w(k) = a_{i_k}.$$

Thus, an integer k in I_r becomes an address or coordinate, and the image of k, given any particular word w, is the letter in w that resides at that coordinate. Note that, by judicious use of the empty symbol e, we can make r as large as we like, or even infinite. In any case, I_r is the formal counterpart of the *absolute space* of Newton.

So we imagine ourselves sitting at a coordinate k in I_r. Given any word

w in $A^{\#}$, we will find ourselves contemplating the letter $w(k) = a_{i_k}$ residing at k. We will call a pair

$$(k, w) \text{ or } (ks, w(k))$$

an *elementary configuration.* Clearly, just as the configuration of a family of Newtonian particles is built from the configurations of each of its members, so too can the configuration of a word w be built in the obvious way out of these elementary configurations.

Now we recall that, in particle mechanics, configuration alone was not enough to give us recursivity. We needed to specify something more, in that case, a notion of *velocity*. Intuitively, velocity actually involves two distinct ideas: a change occurring at the original address and a specification of a new address.

In our present situation, the only kind of change that can occur at an address or coordinate k is the replacement of a symbol originally residing there by another symbol. This in turn means the specification of another mapping

$$\sigma : A \rightarrow A,$$

where for any word w, the composite $\sigma w(k)$ specifies precisely how the alphabet symbol $w(k) = a_{i_k}$, initially residing at coordinate k, is replaced by another symbol $\sigma w(k)$. We note explicitly that, if the new symbol is a blank, the mapping σ amounts to an *erasure;* if the original symbol was a blank, the mapping σ determines an *inscription.*

The set of all such mappings σ is, of course, what we earlier called $H(A, A)$. This set thus plays a role analogous to the *tangent space,* the set of all velocity vectors at a configuration.

Just as in Newtonian mechanics, then, we need to go from a notion of configuration to one of *phase*. This means that, in addition to knowing an elementary configuration (k, w), we must also know σ in $H(A, A)$. The set of all triples of the form

$$(k, w, \sigma)$$

will be *the set of elementary phases*. It is in this set that the relevant notion of *recursiveness* can be introduced.

As we recall, recursiveness is an entailment of "what will happen next" by "what is happening now." "What is happening now" is an elementary phase; what will happen next must be another such phase

$$(k', w', \sigma').$$

Hence, recursiveness is established by positing three rules, three mappings F, g, H such that

$$k' = F(k, w, \sigma),$$
$$w' = G(k, w, \sigma), \quad [7C.1]$$
$$\sigma' = H(ks, w, \sigma).$$

These three rules (F, G, H) are clearly the formal counterpart of the Newtonian idea of (impressed) force. The triple (F, G, H) is what I shall call an *algorithm*.

The first rule in such an algorithm, what I denoted by F above, specifies a new coordinate, a new address, in terms of present elementary phase. Intuitively, it can be thought of as a *shifting of attention* from what is happening at the original coordinate k, to what is happening at a new coordinate k', as specified by the rule F.

As for the second rule G, I will follow convention by positing a specific form for it. Namely, I shall stipulate that

$$\begin{cases} w'(k) = \sigma(w(k)), \\ w'(j) = w(j), j \neq k. \end{cases}$$

This rule is self-explanatory.

Intuitively, the third rule H picks out a new "tangent vector" σ', to be referred to the new coordinate k', and to the symbol $w'(k')$ residing there.

The essence of recursion lies, of course, in the *iteration* of the recursion rules, in this case, the algorithmic entailment of "next phase" from "present phase." The *initial data,* of course, must be specified idependently; in the present case, we require:

1. an initial coordinate k_o
2. an initial word w_o
3. an initial "tangent" σ_o

Successive applications of the algorithm to this initial phase thus generate a corresponding *trajectory,* parameterized by the number of iterations of the algorithm that have been made to reach a particular phase on the trajectory from the initial phase. Hence, as usual, the *time scale* defined thereby is just the set of nonnegative integers. I leave it to the reader, as an exercise, to write out the first few phases on such a trajectory.

It is clear that, as an algorithm pushes us through a trajectory of phases, there will be a corresponding path of words, or spatial patterns of letters, traced out in $A^{\#}$. It is this path that constitutes the main object of interest, in complete analogy with the Newtonian situation.

What I have called an *algorithm* is essentially a Turing machine. The set

$H(A, A)$ of "tangent vectors" constitutes the *set of states* of a Turing machine; the rule H in the algorithm is the *next-state map;* the rule F specifies the *moves* of the machine, etc. A particular word w in $A^{\#}$ constitutes the *input* to the machine, conventionally inscribed, symbol by symbol, on successive squares of an input tape. The algorithm itself, on the other hand, comprises the *reading head* of the machine. I omit the rather tedious and uninteresting details.

I make one observation at this point, which will become important to us soon. It concerns the duality to which I have just referred, between the *reading head,* and the *tape,* which is an essential feature of the Turing machine. The reading head is the *seat of all the entailment,* all the inferential structure, all the mappings. I shall, for obvious reasons, call it *hardware.* On the other hand, the tapes, and anything that can be inscribed on them, contain no entailment at all; I will call them *software.* This distinction between hardware and software has no Newtonian counterpart; the closest we can get to it is to think of particle configurations as "software" and *environment* (the seat of forces) as "hardware." This is clearly an unnatural, and most unattractive, way of trying to talk about physics. Nevertheless, the dualism between hardware and software will, in a rather different way, turn out to be of crucial significance for physics; see section 9B below. This significance rests in turn on the notion of *simulation,* to which we now turn our attention.

7D. Simulation and Programming

The algorithms I introduced in the preceding section are, in a sense, the most general, purely syntactic inferential structures. This point has been endlessly discussed in the relevant literatures, and I will not repeat these discussions here. In any event, if this is so, it follows that a *formalization* is nothing but an alphabet A, the set $A^{\#}$ of words on that alphabet, and a family of algorithms, which constitute its inferential or entailment structure, which produce new words for given words, as I have indicated. Hence, if it were further true that every formal system could be formalized, it would follow that algorithms represent the only kind of inferential or entailment structure that we will ever need.

On the other hand, as I have already discussed at length above (see section 3D), formalization in practice means the replacement of external referents with *equivalent* internal syntactic entailments. Thus, any limits of formalization, i.e., any entailment process that cannot be captured by algorithms, pertain precisely to external referents, and hence, to what is inherently *semantic.*

All this is important to us, because formalisms are what sit on the right-hand side of our diagram (figure 3H.2) above, which in turn is our expression of Natural Law. Hence, if a formalism, whose entailments are in congruence with *causal* entailments in a natural system via such a diagram, can be replaced by an *equivalent* formalization, that would say something quite drastic about causality itself. Specifically, since all entailment in a formalization is *algorithmic,* this fact alone would place profound limitations on the laws of nature, and hence on the things that can sit on the *left-hand side* of the diagram (figure 3H.2).

Since the main thrust of the present work is to put *organisms* on the left-hand side of such a diagram, the limits of formalization are obviously important to us. But as we see, the issue transcends even this and goes to the heart of scientific epistemology itself. The assertion that formalizations suffice in the expression of Natural Law, and hence, that causal entailment is to be reflected entirely in algorithms, is a form of *Church's Thesis,* which I will discuss later (see section 8A below). If it were true, the consequences that follow from its truth would clearly have the most staggering implications for all aspects of human thought. For good or ill, however, it is not true, not even in mathematics itself.

Let us turn now to the question of what algorithms can actually accomplish in a formal context.

Let us begin with some conventional terminology. Suppose we are given an algorithm, a triple (F, G, H), which operates on words w generated from a finite alphabet A. An algorithm is said to *halt* whenever the next phase it produces from a given word w is identical with the present phase. When the algorithm halts, the configuration on which it halts is a word w^*. Thus, the operation of the algorithm can be regarded as establishing a correspondence between words:

$$w \rightarrow w^*.$$

This in turn can be regarded as a *mapping*

$$f: A^\# \rightarrow A^\#$$

defined by

$$f(w) = w^*.$$

Note that, in general, the actual domain of such a mapping f is generally a *subset* of $A^\#$, consisting of all words for which the algorithm halts. Ironically, it turns out that there can be no algorithm for determining whether a given word lies in this domain or not. However, this fact (which in itself already points directly to the limits of formalization) will not enter into the discussion to follow. I simply remark that any mapping in that discussion may be such a *partial mapping,* whose domain need not be all of $A^{\#}$.

Conventional parlance has it that any mapping defined in terms of an algorithm, as I have described, is "recursive." Note, however, that this usage is different from my usage of the term in chapter 3. In our parlance, a mapping (on the integers) was recursive if $f(n)$ entailed $f(n + 1)$ for every n. In that discussion, I gave a necessary and sufficient condition (see [4D.2] above) for such a mapping to be recursive. It is perfectly possible to define mappings f in terms of algorithms, which do not satisfy this condition. Conversely, there is no reason why a mapping recursive in our sense, or rather, the transformation $T f(n) = f(n + 1)$, which generates its values, should be definable in terms of an algorithm. Hence the terms "definable by an algorithm" and "recursive" are not coextensive. Indeed, it will be important for us to carefully differentiate them. Henceforth, I shall call algorithmically definable mappings *simulable,* for reasons to become apparent in a moment (they are also variously called *computable* and *effective*), and reserve the term *recursive* for mappings (simulable or not) satisfying [4D.2] above.

Thus, the word "simulable" becomes synonymous with "evaluable by a Turing machine." In the picturesque language of Turing machines, this means the following: if f is simulable, then there is a Turing machine T such that, for any word w in the domain of f, suitably inscribed on an input tape to T, and for a suitably chosen initial state of T, the machine will halt after a finite number of steps, with $f(w)$ on its output tape.

We shall next do a peculiar thing. Let us take a simulable mapping $f : A^{\#} \rightarrow A^{\#}$, and let us fix a word u in $A^{\#}$. Then we can define a new mapping, $g_u : A^{\#} \rightarrow A^{\#}$, by writing

$$g_u(w) = f(wu) \qquad [7D.1]$$

for all words w in $A^{\#}$.

Let us state this another way. We can trick the mapping f, and the algorithm that computes its values, into evaluating a different map (g_u). This trick is the essence of simulation and of program. In general, we shall say that *f simulates g* if there is a word w in $A^{\#}$ such that $g = g_u$.

What is the relation between u, a word in $A^{\#}$, and g_u, which is a mapping? Let us call the word u the *program* (more accurately, the f/ program) for the mapping g_u. The relation between the program u, and the mapping g_u associated with it, can be thought of as *assigning a referent (g_u) to the word u.* In a certain sense, the program u is a *description* of g_u, recognized by the mapping f.

It is easy to show that, if f is simulable, then so is g_u for any word u.

That means there is an algorithm for computing its values, and hence, a separate Turing machine associated with g_u, whose reading head embodies that algorithm. Hence the program u, which we recall is only a word in $A^{\#}$, can now also be regarded as *a description of that algorithm* and of the Turing machine that embodies it.

Now comes the crucial observation. An algorithm, or a Turing machine's reading head, constitutes *hardware.* It thus constitutes the inferential machinery that *entails* words from other words. The words themselves, on which this machinery operates, are *software.* In particular, *the program u is software.* The simulation of a map g by a map f thus requires an expression of g, or a description of g, as software, as *input.*

It is this fact that gives the simulation relation (the relation between f and g_u in the discussion above) its unique features, features we have not seen before. Namely: the *hardware* of g, the entailment structure that evaluates g, is not put into correspondence with the *hardware* of f, but rather *is imaged in its software;* in *the input to* f. In everything we have seen heretofore, it is precisely the entailment structures that are put into correspondence; that in fact is the very essence of the modeling relation. But in simulation, the *hardware* of the simulator needs to have nothing whatever to do with the *hardware* of what it simulates.

In causal terms, simulation involves the conversion of efficient cause, the hardware of that being simulated, into material cause in the simulator. In essence, this means that one can learn nothing about entailment by looking at a simulation. It also implies that the "time variable," which counts the number of applications of an algorithm to initial data before it halts, is completely unrelated to any corresponding "time variable" in what is being simulated.

It is not too much of an exaggeration to say that the theory of mathematical machines, and its ramifications, is bound up primarily with these basic ideas of programs and simulation. The "general-purpose" machines, the universal simulators, for instance, are essentially those that recognize a maximal number of programs.

We shall not be concerned in what follows with the tactics of simulation; hence we shall not delve too deeply into these issues here. Note, however, the following significant fact. Namely, it is obvious that any machine can simulate itself. Its program for doing so, however, happens to be *the empty word.* Therefore, insofar as we can think of a program as a kind of description of one machine to another, it follows that any machine can have only a trivial description of itself.

I conclude this section with one final remark. The distinction between hardware and software, which I made above, is a distinction between entailment structure, and the syntactic strings of letters on which that structure operates. Once we have introduced the notion of simulation, and the associated notion of program, we can recognize a further distinction, pertaining to software alone: the distinction between *program* and *data.* More precisely, if f simulates g, so that

$$f(wu) = g(w),$$

then the word wu, the argument of f, splits into two distinct parts, with distinct *functions:* the subword u, which is the *program* for g, and the subword w, which is the *argument* for g. This last subword w then constitutes *data.* We note that there is nothing intrinsic to the words themselves upon which such a distinction can be based; it is rather a *relational* distinction, arising from the relation between f and g. Nevertheless, the distinction between program and data is a most important one, when it can be made; indeed, it will assume a central role in some of our further developments.

7E. Simulation and Programming Continued

The considerations of the preceding section lie at the very heart of what constitutes a machine, no matter whether we consider "machine" to connote a purely formal object or a kind of natural system in the ambience. This fact alone would justify the most careful discussion of the basic concepts of simulation, and of programming, which we have introduced. In addition, however, the idea of simulation has repeatedly been confounded with modeling, a confounding that has badly obscured both concepts. Accordingly, I devote the present section to an exploration of the deeper strata of entailment on which simulation and programming rest.

It might be thought that our discussion of the preceding section is the most general setting possible; what could be more general than starting with an alphabet of undefined symbols and working syntactically up from there? Nevertheless, there are wider formal contexts in which the ideas I have introduced can be seen much more clearly, in which the subtle, and apparently arbitrary, interplay between syntactic operations and the arguments on which they operate assume a more tangible and apprehensible substance. In the present section, I attempt to exhibit this substance.

To begin, note that the set $A^{\#}$, which was the focus of our treatment of simulation and programming in the preceding section, is not an arbitrary set. On the contrary, it manifests a great deal of internal structure, which we used heavily. It is canonically generated from a finite alphabet A and manifests a distinguished binary operation (concatenation); essentially, we used this operation *in* $A^{\#}$ to create new operations *on* $A^{\#}$; conversely, we saw how to express operations *on* $A^{\#}$ in terms of specific words or programs *in* $A^{\#}$. (Clearly the spirit of Gödel can be perceived at this point, hovering nearby.) I referred to these two aspects as *hardware* (operations *on* $A^{\#}$) and *software* (elements *in* $A^{\#}$), respectively, and showed how programming, and simulation, involved the expression of the one in terms of the other. I now tease these ideas apart still further, primarily by pulling concatenation away, removing its intrinsic character, and treating instead an arbitrary binary operation on an arbitrary set.

Accordingly, let us suppose that X is an arbitrary set and that we have specified somehow a distinguished mapping

$$\sigma : X \times X \rightarrow X$$

This map σ will play the role of concatenation in $A^{\#}$ but now in a completely general setting. If x is an element of X, and if we can write

$$x = \sigma(\xi_1, \xi_2),$$

then it is natural to regard the elements ξ_1, ξ_2 in X as constituting a *factorization* of x, with ξ_1, ξ_2 as *factors.*

Suppose we now fix an element u in X and look at the set of all pairs of the form

$$(\xi, u),\ \xi \text{ in } X. \qquad [7E.1]$$

Applying σ to each such pair clearly defines a subset of X, the subset of all elements x that possess the fixed element u as a factor. This subset is in fact simply the image under σ of the set,

$$\mathrm{X} \times \{\mathrm{U}\},$$

which in turn is a subset of X $\times$ X.

In addition, whenever we have a Cartesian product, we have also the *natural projections*. At the moment, we are interested only in the one defined by

$$\pi(\xi_1, \xi_2) = \xi_1,$$

which is itself a kind of binary operation,

$$\Pi : X \times X \rightarrow X.$$

Consider now the diagram

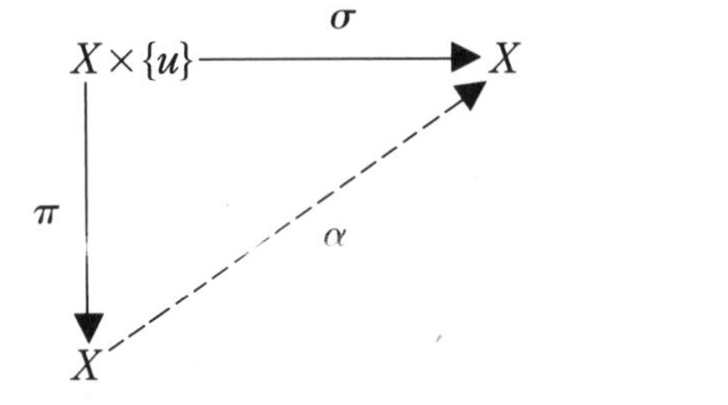

[7E.2]

If factorization is *unique* (i.e., $\sigma(x_1, u) = \sigma(x_2, u) \Rightarrow x_1 = x_2$), then there is clearly a unique mapping $\alpha : X \rightarrow X$, indicated by the dotted arrow, *entailed* by σ, by χ, and by the choice of u, which makes the diagram commute.

So far, we have not said very much. But now suppose we take another mapping $f: X \rightarrow Y$, where Y is just some other set (possibly X itself), and attach it to the diagram as shown:

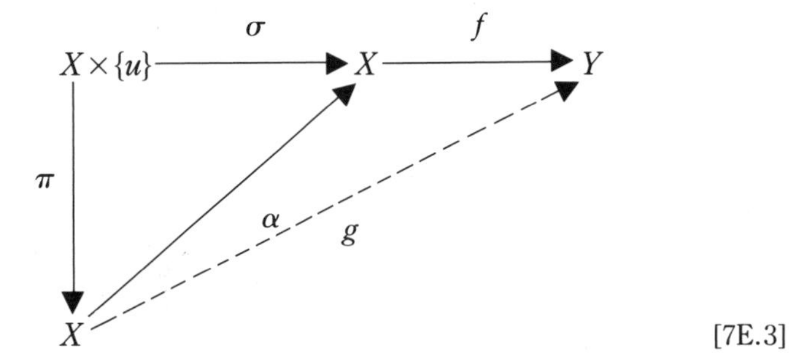

[7E.3]

There is then a mapping $g: X \rightarrow Y$, *defined* by

$$g = f\alpha, \quad [7E.4]$$

which makes the whole diagram commute. The relation between f and g is what we have seen before; we may say that f *simulates* g. Moreover, we have also seen the relation between g and the fixed element u before; we may call u the *program* for g, given f.

Turning all this around, we can say that a map g is *simulable* by a map f precisely when we can find an element u in X for which the commutative diagram above can be constructed. Commutativity here means precisely that

$$g\, \pi\, (\xi, u) = f\, \sigma\, (\xi, u) \quad [7E.5]$$

for every element ξ in X.

We can now discourse at length upon the simulation relation between g and f, as embodied in the diagram [7E.3]. This is in fact a complicated diagram; its complication rests on the fact that the set X appears three times in the diagram, and in each appearance, *it plays a different functional role;* it *connotes a different thing.* The failure to distinguish between these distinct connotations, which is rendered easy because it is the *same set X* that appears in these distinct rules, has wrapped the entire concept of simulation in clouds of needless obscurity.

Indeed, we may remark at the outset that the diagram [7E.3] in itself represents a specific kind of relational system, as I have previously described (see section 5H above). It is the whole diagram, the whole relational system, that endows each of its separate elements with a specific functional role and that in this case confers a specific relation (simulation) upon the two mappings f and g that appear in it. This fact should be borne in mind as we proceed with our analysis.

To look more closely at these distinct functional roles played by the set X, let us redraw the diagram [7E.3], with the different occurrences of X labeled as shown:

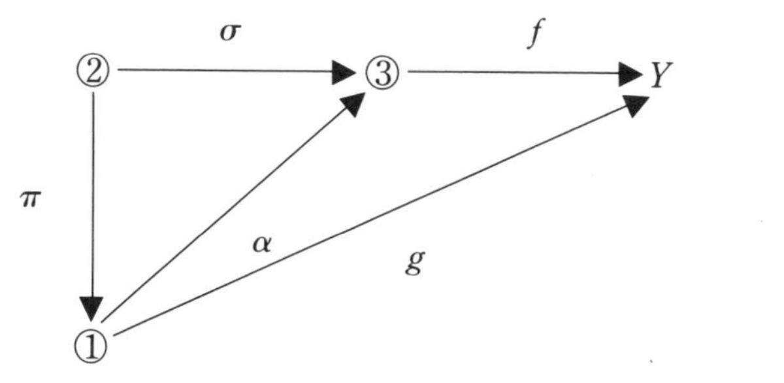

[7E.6]

In the position ①, the set X comprises the *set of arguments for the mapping g which is being simulated.* In the position ②, each such argument x has been lifted into a unique pair (x, u), where now u in X plays the role of the *program* for g. In position ③, X is the range of the binary operation σ and now constitutes *the set of arguments for the simulator f of g.*

The net effect of all these circumlocutions is *to pull the mapping g inside the set X itself,* to give g a representation in terms of its own arguments. Such an apparently paradoxical situation can arise only in very special circumstances and requires the multiple significations with which we have endowed the set X and its elements. In such circumstances, the actual relation between f, the simulator, and g, that which is being simulated, is not at all what it appears to be. Namely, if we just look at the piece of the diagram [7E.3] indicated below:

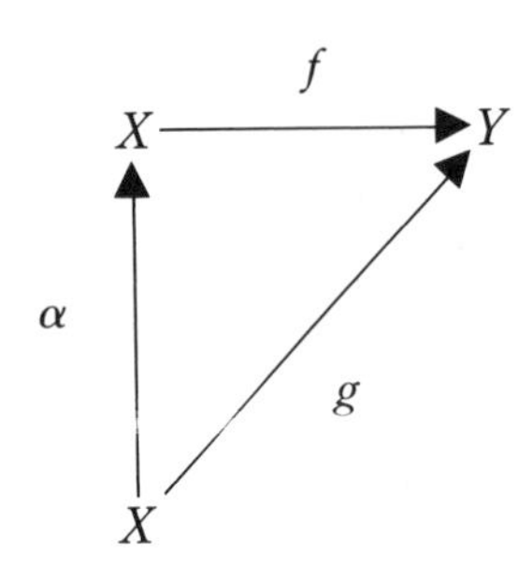

it appears that the mapping g is simply factorized through the simulator f, via the mapping α (see [7E.4] above). But looking at the entire diagram, we see that this map α is really only the binary operations σ in disguise and that what looks like a factorization of *mappings* is actually a factorization of *elements* of X Indeed, α is completely independent of f and g; it depends *only on* σ.

We can thus begin to see how special the concept of simulation really is. We will see this even more graphically later (see section 9B et seq. below). The main feature to which I draw attention here is that mapping α, which looks at first like a simple encoding of the domain of g into the domain of f, is not open to choice; it is *entailed* by the (syntactical) structure in X alone, independent of any mappings.

I shall now contrast this notion of simulation with that of a model. This is the last ingredient we shall need in coming to grips with the general concept of the machine.

7F. Simulations and Models

The concept of simulation provides us with a new and different way of comparing two formalisms, different from the modeling relation we introduced earlier. In the present section, I compare, and contrast, these two modes of comparison. This comparison is interesting and important in itself; but more to the present purpose, both modeling and simulation will enter essentially, and in different ways, into our characterization of machines and mechanisms.

Let us refresh our recollection of modeling relations, as manifested between two formalisms (and in particular, between a formalism and a formalization). We recall the characteristic commutative diagram (figure 7F.1) that brings the inferential structures (schematically represented by the arrows ① and ③) of the two formalisms into congruence, through

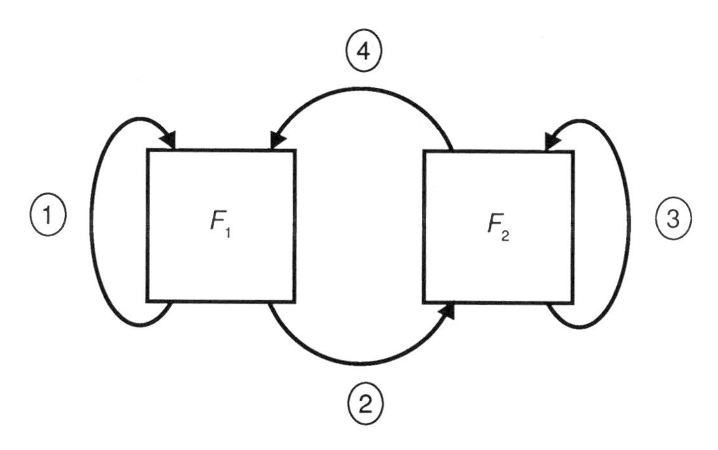

FIGURE 7F.1

suitable encoding and decoding between what these inferential structures act on.

When such a congruence between inferential structures is established, it amounts to *an extension of the encoding and decoding arrows (② and ④) between the two formalisms to their inferential structures themselves.* That is, the commutativity of the diagram allows us to encode the arrow ① into the arrow ③, and decode the arrow ③ into the arrow ①. All entailment relations *within* the formalisms are thus preserved by the modeling relation *between* them; that is precisely why we can learn about one of them by studying the other.

Let us further recall that, at the level of the formalisms F_1, F_2 themselves, *the inferential structures that define them are themselves unentailed.* They are simply presumed given at the outset, and we merely study their consequences. Since modeling relations can introduce no new entailments in themselves, *they also preserve the absence of entailment* of the inferential structures they compare. Using a picturesque and appropriate causal language, these inferential structures act as first causes; being themselves unentailed within the formalisms they define, they cannot be effects of anything in those formalisms, but they generate effects (consequents) through the entailments they themselves establish.

We proceed by contrasting a *model* of a mapping with a *simulation* of that same mapping.

Let us refer again to the diagram exhibited in figure 7F.1 above, where F_1 now constitutes a map to be simulated, and F_2 is to be the simulator. We shall think of the map to be simulated as the arrow ① in that diagram; the simulator as the arrow ③. As with modeling, there must be an encoding and a decoding. It is precisely here, however, that the basic difference arises.

Namely, in simulation, the arrow ①, the map to be simulated, is pulled *inside* the box F_2 by the encoding itself. Indeed, *the entire left side of the diagram now disappears* inside the right-hand box. Consequently, the inferential machinery of the simulator F_2 has nothing in F_1 to decode into, nothing in F_1 to which it can correspond. Simulation is thus not a congruence between inferential structures.

Since this distinction between modeling and simulation is so important, I shall describe it in a number of other terminologies. Let me first speak of it in terms of what happens to entailment. To avoid confusion with our earlier discussions, I denote the map to be simulated by $\varphi : X \to Y$. This, as we know, amounts to the entailment

$$\varphi \Rightarrow (x \Rightarrow \varphi(x)). \qquad [7F.1]$$

A modeling relation preserves this entailment:

$$\alpha(\varphi \Rightarrow (\alpha(\varphi(x)) \qquad [7F.2]$$

where α is the encoding arrow ②. In each case, φ and its model are both themselves unentailed.

But in a simulation ψ, we have a completely different situation: namely,

$$\begin{cases} \psi \Rightarrow \{[\alpha(\varphi),\ \alpha(x)] \Rightarrow \psi[\alpha(\varphi),\ \alpha(x)]\} \\ \qquad \psi[\alpha(\varphi),\ \alpha(x)] \ = \ \alpha(\varphi(x)). \end{cases} \qquad [7F.3]$$

From this we can see most clearly that, instead of a *congruence* between two different inferential structures φ and ψ, themselves each unentailed, simulation gives us an *entailment,* by the simulator, of the entailment φ it simulates. In other words, simulation turns φ into an effect, a consequence of the simulator ψ, which may have no relation to φ at all.

We thus in general can learn nothing about φ by looking at a simulation of it. We can see this most clearly by looking at a situation in which two different mappings φ, φ' are simulated by the same simulator ψ. Such a situation superficially resembles the diagram exhibited in [3H.3] above, which we have discussed in connection with modeling relations. But whereas the possession of a common *model* imposes a fundamental relation (analogy) between the systems themselves, the possession of a common simulator tells us absolutely nothing about them beyond that fact.

Indeed, readers may find it illuminating to go back to the discussion of section 3H above, replacing "model" by "simulation," and see how far they get. Such an exercise exhibits, in the most graphic way possible, the difference between the two concepts.

We can express this difference in yet one more way, in terms of the distinction I made above between "hardware" and "software." Insofar as such a distinction is intrinsically meaningful, it is always preserved by modeling relations. It is never preserved by simulations.

Chapter 8

Machines and Mechanisms

WE NOW have all the basic ideas we need in our hands; things are going to happen rather quickly from now on.

8A. Review

In chapter 6 I introduced the notion of the category $\boldsymbol{C}(N)$, the category of all models of a system N. I showed that this category has a good deal of structure and that this structure itself tells us metaphorically about N. For instance, $\boldsymbol{C}(N)$ has a natural partial order between its objects (i.e., between the models of N); one model M_1 is bigger than another model M_2 if and only if M_2 can be encoded into M_1 but not conversely. In a purely formal setting, the ordering in question reflects the refinements of the equivalence relation corresponding to M_2 by that corresponding to M_1. We also saw that, given any pair of models in $\boldsymbol{C}(N)$, we could put them together to get a new one, generally bigger than either of them.

This much is all generally true. We also saw an essential distinction between two *kinds* of models of N, which we called analytic and synthetic models. Intuitively, analytic models were based on the notion of *observables* and led to encodings of N into Cartesian or direct products of spectra of observables. Synthetic models, on the other hand, were based on the idea of direct sums of disjoint summands. As we noted, the two are in general very different, but as models, they could be compared in $\boldsymbol{C}(S)$ relative to the ordering relation in $\boldsymbol{C}(N)$, and they could be combined. In particular, I suggested that the existence and distribution of synthetic models in $\boldsymbol{C}(N)$ reflected something important about N itself. The present chapter is devoted in large part to an exploration of what it is about N that synthetic models actually do reflect.

In chapter 7, I introduced the idea of a *simulation* of a formalism.

Roughly speaking, I showed that a formalism can be simulated if its inferential structure could be expressed as software to a mathematical machine, in particular, as *program.* As I suggested, this places severe restrictions on that inferential structure; simulability of a formalism is a strong condition to be mandated of it.

We are now in a position to put these apparently unrelated ideas together and see what happens.

8B. Machines and Mechanisms

As we have seen, given a natural system N, we have formalisms F associated with it as its models, simply by virtue of Natural Law itself. We now also have a condition (simulability) that may be mandated of formalisms. Putting the two together defines for us a *class* of natural systems, those whose models, as formalisms, satisfy that condition.

Let us give a name to this class. We shall say that *a natural system N is a mechanism if and only if all of its models are simulable.*

We shall further say that *a natural system N is a machine if and only if it is a mechanism, such that at least one of its models is already a mathematical machine.*

On the face of it, these seem peculiar ways of characterizing mechanisms and machines from among the class of natural systems. But this peculiarity stems only from my expression of these concepts in terms of the models of N, rather than try to talk directly about N itself. This is all that Natural Law entitles us to do. We have so far in this volume nothing that transcends those entitlements, and I shall do nothing in what follows that transcends them; on the other hand, it is my aim to utilize precisely these entitlements to the full extent. It is my main contention, in fact, that contemporary science, as a whole, does not do this; by the time we are done, this fact and its consequences will be quite apparent.

In fact, my characterization of mechanism will be seen to be nothing but Church's Thesis (see section 7D above), explicitly and fully manifested in its true material garb. As I noted earlier, the intent of Church's Thesis was initially to characterize vague notions of "effective" calculability and algorithm, and hence programmability in Turing machines. But because it is so easy to equivocate on the word "machine" and because everything that happens in the material world must certainly be considered "effective," Church's Thesis has always been tacitly supposed to have a physical content as well. For instance, consider these words, taken from Martin Davis' book *Computability and Unsolvability* (New York: McGraw-Hill, 1958):

> For how can we ever exclude the possibility of being presented someday (perhaps by some extraterrestrial visitors), with a (perhaps extremely complex) device or "oracle" that "computes" a noncomputable function? (p. 11)

My definition of mechanism above merely characterizes the world in which this equivocation is legitimate (i.e., within which Church's Thesis becomes elevated to a Law of Nature) and investigates the consequences.

In any case, this characterization of mechanisms and machines will be seen to be completely intrinsic and unrelated to any questions of history, or of fabrication, by which material machines are usually characterized. This very fact will become important later (see chapter 10), because it make it manifest that questions pertaining to fabrication are of an entirely different order from those we are now considering.

Furthermore, in this context, we can see the true import of additional suppositions, such as

Every natural system is a mechanism,

or

Every organism is a machine.

These are, of course, part of the very fabric of which contemporary science is composed. We are now in a position to assess them directly; this has been in fact my intention all along.

Let me then simply state, at this point, some of the conclusions that will be established in the remainder of the present chapter. These are:

Conclusion 1: If a natural system N is a mechanism, then it has a unique largest model $M^{\max}$. That is, the category $\boldsymbol{C}(N)$ contains a unique maximal element, with respect to its natural partial ordering. Epistemologically, this model contains everything knowable about N, according to Natural Law.

Conclusion 2: If a natural system N is a mechanism, then there is a (necessarily finite) set of minimal models $M_i^{\min}$.

Conclusion 3: The maximal model $M^{\max}$ is equivalent to the direct sum of the minimal ones;

$$M^{\max} = \sum_i M_i^{\min}$$

and is therefore a synthetic model.

These conclusions will allow us to define *states* for the maximal model, in terms of the minimal ones, in a completely intrinsic way, so that change of state is necessarily recursive in $M^{\max}$ if they are in the $M_i^{\min}$. On the basis of this, we can then conclude further:

Conclusion 4: If N is a mechanism, then analytic and synthetic coincide in $\mathbf{C}(N)$; direct sum = direct product.

Finally, I shall introduce a notion of fractionability, relating properties of N to corresponding properties of its synthetic models. We will then have

Conclusion 5: Every property of N is fractionable.

The sum of these conclusions can be expressed very simply: a mechanism is a completely syntactic kind of thing. Intuitively, the syntactic nature of mechanisms is their salient characteristic. It appears here as a consequence of imposing simulability on the category of models; in retrospect, we see that our unusual characterization of mechanism in terms of simulability was in fact a natural one.

Let us now proceed to establish these conclusions.

8C. On the Largest Model of a Mechanism

To show that a mechanism must have a unique largest model is very easy. Let us begin by fixing an appropriate mathematical machine, say, a universal Turing machine. To say that a model (or indeed, any formalism) is simulable means that there is a program for it, i.e., a finite word built of the alphabet letters of the machine.

Now let us suppose that N is a mechanism (every model simulable) but that $C(N)$ contains no largest model. Then we can find an infinite sequence of increasingly refined models

$$M_1 < M_2 < M_3 \ldots < M_i < \ldots$$

in $\mathbf{C}(N)$. Each of them must have a program of finite length; we thus get a corresponding sequence of words $\{u_i\}$ in $A^{\#}$.

But now we can form the intersection M of all the models M_i. This is again a model of N, and unless the sequence $\{M_i\}$ already terminates after a finite number of iterations (i.e., unless there is an n_o such that $M_n = M_{n_o}$ for $n > n_o$), this M is strictly larger than any of the M_i;

$$M_i < M, \text{ all } i$$

But then, by hypothesis, M must have a program u as well. We thus end up with a countable family of distinct programs, each of which is a distinct word of finite length on a finite alphabet. This is clearly impossible.

Moreover, this largest model must be unique. For if there were another such largest model, say M', and if M' were not the same as M, then the two would have to be incomparable (i.e., we could have neither

$$M < M' \text{ or } M' < M$$

without violating our hypotheses). In that case, we can take their mutual refinement, which is again a model and dominates them both, again contradicting our hypothesis. QED.

My deceptively simple argument actually demonstrates much more than I am claiming. For instance, we can see immediately that, if N is a mechanism, there are only a finite number of refinements between any model and the unique maximal one. But I shall not need these corollaries, and hence, I do not pursue them here.

8D. On the Smallest Models of a Mechanism

Exactly the same kind of argument as I have just employed, only run the other way, establishes the existence of minimal or smallest models. There is of course no uniqueness this time. Nevertheless, it is clear that the totality of such minimal models of N must constitute only a finite set, and the elements of this set are all mutually incomparable. I will leave the easy details as an exercise for the reader.

8E. Maximum Model From Minimal Models

We have seen in the preceding sections that, if N is a mechanism, then $\mathbf{C}(N)$ contains a unique largest model $M^{\max}$ and a finite family $\{M_i^{\min}\}$ of mutually incomparable minimal models. I shall now establish a powerful relation between $M^{\max}$ and the $M_i^{\min}$. Namely, I shall show that

$$M^{\max} = \sum_i M_i^{\min};$$

the largest model is the direct sum of the minimal ones.

Once again, the argument is now very simple, for the direct sum of the minimal models is again clearly a model. If it is not already $M^{\max}$, then it is certainly refined by $M^{\max}$. The proof of my assertion will consist in showing that there must then be another model P in $C(N)$, such that (1) the mutual refinement of P and $\Sigma M_i^{\min}$ is $M^{\max}$, and (2) this P must also be minimal, or else a direct sum of minimal models, different from all the $M_i^{\min}$. This will, of course, contradict the hypothesis that the $M_i^{\min}$ are *all* the minimal models of $\mathbf{C}(N)$; hence P must be vacuous, and the conclusion will follow.

This argument rests on the fact that, if R is an equivalence relation, and R defines another equivalence relation R_1, so that

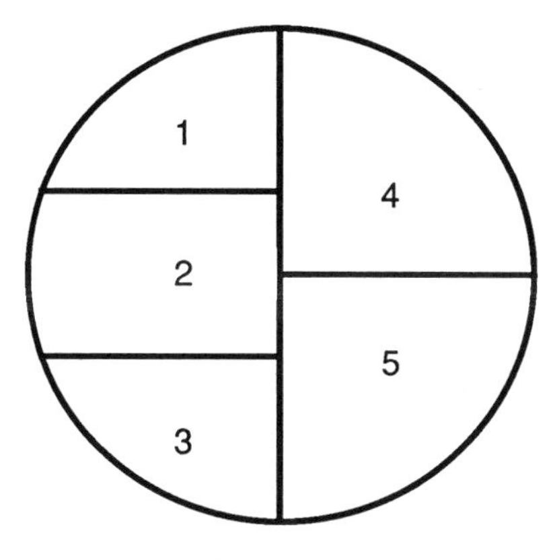

FIGURE 8E.1

$$R_1 < R,$$

then the equation

$$R_1 \cap X = R \qquad [8E.1]$$

can be solved (generally in many ways) for X; i.e., there is another equivalence relation X, such that R is the common refinement of R_1 and X.

As with most results of this kind, the general proof is not difficult, but it is tedious and uninformative. So I will content myself with exhibiting the main idea of the argument and leave the details as an exercise for the reader. For instance, suppose we visualize R as consisting of, say, five equivalence classes as indicated in figure 8E.1, and suppose that R_1 consists of the two classes indicated in figure 8E.2. Here we have

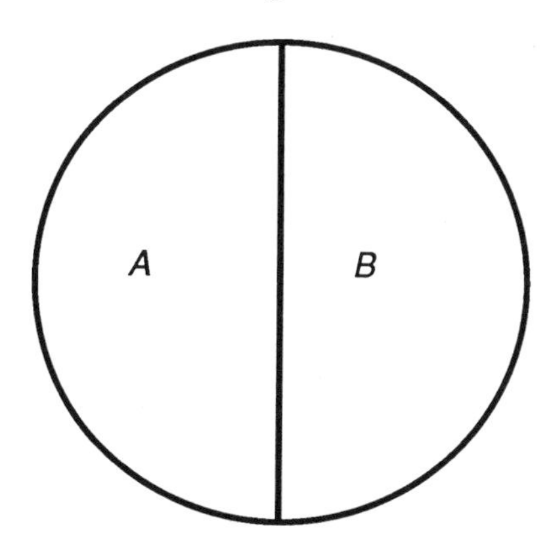

FIGURE 8E.2

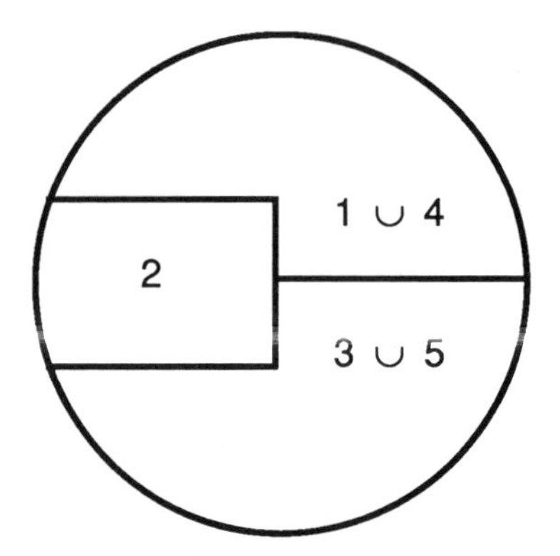

FIGURE 8E.3

$$A = 1 \cup 2 \cup 3,$$
$$B = 4 \cup 5.$$

Then one solution of [eq. 8E.1] is given by the diagram shown in figure 8E.3. As will be seen, the trick is simply to rummage through the classes of R and their unions and select from them an appropriate subset that constitutes the classes of a new equivalence relation X (which R necessarily refines), satisfying [eq. 8E.1]. It should be obvious that this can always be done, and hence, our little example is really perfectly general.

Returning now to the main line of the argument, we see that if R^{max} refines $\sum_i R_i^{min}$ without being equivalent to it; i.e., if

$$M_i^{min} < M^{max},$$

then there is indeed another model P in $C(N)$, such that

$$P \cap \sum_i M_i^{min} = M^{max}.$$

Using the same argument we did before, the hypothesis that N is a mechanism implies that either P is itself a minimal model or else is composed of minimal pieces. As I said, it is evident that these minimal pieces are different from all the M_i^{min}, and this contradicts our hypotheses.

From these results, we can already begin to see how strong, and hence, how restrictive, the hypothesis that N is a mechanism really is. In terms of the category $C(N)$ of all models of N, we see that N must have a unique largest model and a finite family of smallest models. Epistemologically, Natural Law says that all we can know about N is already manifested in that largest model. The result of the present section says that, in these terms, the relation between the largest model and the smallest ones is a *purely syntactic relation;* i.e., it says that all we can know about N inheres already in (1) its spectrum of smallest models and (2) the purely syntactic rules, independent of N, for generating direct sums in $\mathbf{C}(N)$.

8F. On States and Recursivity in Mechanisms

One corollary of the purely syntactic relation that exists between $M^{\max}$ and the $M_i^{\min}$ is the following: if the latter have states, then we can use these to assign states to $M^{\max}$ in a unique way. This fact is indeed inherent in the very idea of a direct sum; see section 6D above. This is merely another way of stating what I said earlier; once we have encoded N into the $M_i^{\min}$, a corresponding encoding of $M^{\max}$ is thereby *entailed.*

A state of $M^{\max}$ is thus built up syntactically from, or coordinatized by, the corresponding states of the $M_i^{\min}$. Hence, by definition, we can express everything about $M^{\max}$ in terms of these coordinates; in terms of the way the $M_i^{\min}$ are injected into $M^{\max}$. Furthermore, according to Natural Law, $M^{\max}$ already embodies everything we can know about the mechanism N that it models. Therefore, everything we can know about N is also expressible in terms of these coordinates.

It remains to say a word about *recursion* in mechanisms. The discussion is not completely automatic, because recursion (i.e., the entailment of $f(n + 1)$ by $f(n)$) is not in general coextensive with simulability (which has to do in general with the entailment of $f(n)$ from n); see section 7D above. Once again, however, the presence of a largest model $M^{\max}$ of a mechanism comes to the rescue.

If x represents some magnitude pertaining to a mechanism (e.g., a state variable), then as we have abundantly seen (see section 4H), its temporal rate of change $\dot{x}$, its velocity, is also. Likewise its acceleration $\ddot{x}$, and indeed, all of its infinitely many higher temporal derivatives $x^{(n)}$. These must all obviously encode into the largest model $M^{\max}$. If these were all *independent* (i.e., unconstrained, see section 3F above), then in general we would need to supply all of their synchronic values as *software* (data) to whatever mathematical machine *simulates* $M^{\max}$; such a simulator exists by definition.

But this is impossible in a simulator, which requires finiteness in its software, a fact I have used (section 8B above) in connection with *program* rather than *data.* Accordingly, a mechanism is inconsistent with the independence of all the temporal derivatives $\{x^{(n)}\}$ of any quantity pertaining to it. Hence there must exist *relations* between their values at any instant; in a word, the mechanism must in this sense be *constrained*. But as we have seen above (see section 4F), this means precisely that the mechanism is governed by the differential equations that express these constraints. This embodies the recursion by virtue of which *present state* entails *next state* in a mechanism.

We conclude then that, in a mechanism, simulability actually entails recursion in M^{max}. Stated another way, the constraints mandated by simulability are manifested in the form of differential equations, and differential equations are by their very nature recursive.

I may note, although I cannot discuss these matters here in any detail, that it is precisely here where the encodings typical of quantum mechanics differ from those of classical. In some sense, quantum mechanics has to deal with precisely the independence of all the $x^{(n)}(t)$. Hence, if it is to preserve recursiveness (which it identifies with causality) it *cannot* encode "states" of a system into the $\{x^{(n)}\}$ *directly*. In effect, it gives up recursiveness *on these quantities*, in exchange for manifesting it elsewhere (via the Schrödinger equation). This is primarily why the encodings characteristic of classical and quantum mechanics are incompatible. The differences between them are of great significance and should be kept in mind when we come to discuss complex systems (see section 9F below). But despite all these differences in detail, quantum mechanics remains mechanics.

8G. Synthesis and Antisynthesis: Fractionability

As we have seen, a mechanism N necessarily has a largest model M^{max}, which in turn is built up syntactically as a direct sum of minimal models M_i^{min}. It is thus necessarily a synthetic model, one that possesses a definite coordinatization inherited from the M_i^{min}, as we have seen. Also, since M^{max} is the maximal model, it encodes everything about N that we can know. Thus, putting the two together, everything we can know about N can be expressed in terms of that mandated coordinatization.

On the other hand, Natural Law also tells us that, whatever we can know about N is itself embodied in *some* model M belonging to $\mathbf{C}(N)$. We shall now be concerned with the relation between such an arbitrary model M in $\mathbf{C}(N)$, the M_i^{min}, and M^{max}. As we shall now see, if N is a mechanism, these relations are very sharply circumscribed indeed.

Obviously, since M^{max} is the largest (i.e., most refined) model of N, we must have

$$M < M^{max},$$

M^{max} must refine M. Stated another way, M must be expressible as a quotient of M^{max}; i.e., we can write

$$M = M^{max}/R,$$

where R is some suitable equivalence relation on M^{max} itself.

We shall now show that

$$M = M^{max}/R = \sum_k M_{i_k}^{min}. \qquad [8G.1]$$

That is: an *arbitrary* model M of a mechanism N is equivalent to a direct sum of minimal models.

This conclusion can be established in a variety of ways. Perhaps the most elegant invokes an argument I have already used, when I showed (see section 8E above) that M^{max} itself was the direct sum of minimal models.

Suppose then that M is a model in $C(N)$. Then M is certainly refined by a direct sum of minimal models, e.g., by M^{max} itself. Take a smallest such synthetic refinement; call it

$$M^* = \sum_k M_{i_k}^{min}.$$

If $M \neq M^*$, then we can find another model X satisfying the condition

$$M \cap X = M^*,$$

which, as before, would violate our hypotheses. Hence $M = M^*$.

What I have shown is that, if N is a mechanism, then *any* model is equivalent to a synthetic model. Hence in particular, any analytic model must be a synthetic model. As we have seen above (see section 6E), we cannot expect this to be generally true; it amounts to requiring that direct sums and direct products coincide in $\mathbf{C}(N)$. The fact that it *is* true if N is a mechanism shows, on the one hand, how special N itself must be; on the other hand, it accounts for the attractiveness of restricting attention to mechanisms in the first place.

Two aspects of this situation are worth noting specifically. The first is embodied in [eq. 8G.1] and can be stated as follows: if N is a mechanism, then any property or aspect of N that can be embodied in a model M can be *localized,* in the sense that it can be expressed as a property of a direct summand of M^{max}. That is, we can write, from [eq. 8G.1],

$$\begin{aligned} M^{max} &= \sum_k M_{i_k}^{min} + \sum_{j \neq i_k} M_j^{min} \\ &= M + \sum_j M_j^{min}. \end{aligned}$$

One way of stating this conclusion is as follows: given any property of a mechanism N, embodied in a model $M \neq M^{max}$, that property is already *localized* in a direct summand. We shall say that N can be *fractionated,* i.e., separated into two parts modeled by disjoint direct summands, such that the property in question is manifested in one of those parts.

Hence, if N is a mechanism, then either any such property is fractionable in this sense or else any model of it is equivalent to M^{max}. This notion of fractionability is going to become very important to us in our subsequent work, because in some sense it characterizes, by itself, the concept of a mechanism.

My second remark is really an outgrowth, or generalization, of what I have already said. Namely, if N is a mechanism, then there is really only one mode of *analysis* of N, a mode that we may call *antisynthesis.* As we have seen, any mode of analysis of a natural system N culminates in models of N. If N is additionally a mechanism, we have just seen that the resulting model M must be equivalent to a synthetic one, which serves precisely to localize the fruits of our analysis in a direct summand of M^{max}. Hence, if N is a mechanism, we can get at any analytic model M *either* through direct synthesis (i.e., by taking a direct sum of appropriate minimal models of N), *or* by throwing away direct summands of M^{max}. This last process, which involves the pruning of M^{max} by discarding some of its direct summands, is clearly an inversion of the synthetic process by which M^{max} is generated from the M_i^{min}; accordingly, it is reasonable to call it *antisynthesis.*

Thus, if N is a mechanism, any mode of analysis whatsoever is equivalent to a process of antisynthesis. Stated another way: in a mechanism, *analysis coincides with antisynthesis.* This special feature of mechanisms will also be important to us later. For instance, when we come to talk about those even further specialized mechanisms we shall call machines, some of the localized parts arising from fractionability will be the bearers of *functions* in the machine; accordingly, *such functions are always localized into corresponding organs.* Furthermore, and even deeper, is the essential identity between *reductionism and antisynthesis;* the two are indeed essentially the same. Hence in particular, if a natural system possesses a model inequivalent to any synthetic model (which is one way it can fail to be a mechanism at all), reductionism is useless.

8H. Mechanisms and Contemporary Physics

As we have seen, a contemporary physicist will feel very much at home in the world of mechanisms. We have quite deliberately created this world without making any physical hypotheses, beyond requiring the simulability of every model. We have thus put ourselves in a position to do a great deal of physics, without having had to know any physics. That fact alone should indicate just how special the concept of mechanism really is. It is my contention that contemporary physics has actually locked itself into this world; this has of course enabled it to say much about the (very special) systems in that world, and nothing at all about what is outside. Indeed, the claim that there is nothing outside (i.e., that every natural system is a mechanism) is the sole support of contemporary physics' claim to universality.

We have seen in ample detail what this world of mechanisms is like. If N is a mechanism, then its category of models, $\mathbf{C}(N)$, must have a very special structure. There must be minimal models in it, a finite number of them, which, if we like, we can call (elementary) particles. These minimal models come equipped with states; it does not matter whether we call them phases or wave functions or whatever else we care to call them.

There is also a maximal model $M^{\max}$, which can be built up synthetically by purely syntactic operations (i.e., by direct summations) from the minimal ones. It is thereby coordinatized by the minimal ones; it inherits a fixed state structure thereby, and with it, a mandated or entailed encoding from N itself. Moreover, in $\mathbf{C}(N)$, there is a way of expressing any model (particularly, analytic models or direct products) in terms of the syntactic operations of forming direct sums; in that sense, direct products and direct sums coincide. We called this fractionation or antisynthesis above. This in turn mandates a lot of linearity, in $M^{\max}$, and in $\mathbf{C}(N)$ itself.

If $M^{\max}$ is recursive (i.e., manifests enough entailment to make next state be entailed by present state), then all the considerations of chapter 4 are immediately applicable. In particular, we have Taylor's Theorem in some form, and we have the constraints that in turn transmute into the equations of motion, which precisely serve to express how next state is entailed by present state. On the other hand, this is the sum total of the entailment we can expect in a mechanism.

In short, we see the full form of contemporary physics flowing solely and entirely out of the hypothesis of mechanism. The only things that do not follow from that hypothesis are (1) that there is a universal finite family of minimal models, corresponding to a fixed, finite number of different kinds of "elementary particles," which suffice for *any* natural system N, and (2) the assignment of states to such "elementary particles." And it so happens that both of these remain matters of complete uncertainty and hence controversy in contemporary physics itself. As to (1), we have shown only that, given a particular mechanism N, its maximal model can be constituted syntactically from a finite family of minimal models; it needs another hypothesis to suppose that two different mechanisms N_1, N_2 are constituted from the same set of minimal models. And indeed, the nature and number of "elementary particles" is constantly changing, both in experimental and in theoretical physics. And as to (2), we need only observe that, as (1) changes, so does (2).

Note also that these problems (1) and (2) arise entirely within the world of mechanisms. However we want to answer them, the answers cannot take us outside that world. In particular, they cannot give us new modes of entailment. They can only reexpress, or subdivide, or play other such syntactical games, with whatever entailments come directly from the hypothesis of mechanism. And as we have seen, this is not very much.

Chapter 9

Relational Theory of Machines

ARMED AS we now are with all of the special structure that pertains to the concept of a mechanism, we will turn our full attention to the particular class of mechanisms we called *machines.* As we recall, a machine is a mechanism, one of whose *models* is a mathematical machine. Hence, our concern is with machines as natural systems, machines as objects of scientific study.

9A. Machines

Our first result will be to show explicitly that machines in general admit *relational descriptions.* We will in fact see in detail how the relational descriptions of machines arise from, and are related to, their underlying "physics" as we have developed it in the preceding chapter. Once we have done this, it will become transparent how these relational descriptions can be detached from that "physics." As such, we obtain a formal encoding of pure organization, as a thing in itself.

Because machines are so special, we will find that the organizations they can manifest are also special. The culmination of the present chapter will be a theorem precisely characterizing their limitations and a discussion of their significance for biology, for physics, for technology, and even for mathematics itself.

My treatment will be based on the distinction between *hardware and software,* which as we have seen above (see section 7E et seq.) is perhaps the essence of the concept of a mathematical machine. Using the hypothesis that such a machine is a model, I will pull this distinction back into the largest model, which exists by virtue of the fact that a machine is a fortiori a mechanism. As we shall see in detail, we are then enabled to decompose this largest model into a number of disjoint functional units or operators (comprising the hardware) and that on which they operate (the software). The functional units will be called *components,* see section 5D above; their specific action on software will be expressed in terms of *inputs* to these components, and the entailment of corresponding *outputs.*

The result will be to express the overall behavior of the machine as a family of *ternary* relations between inputs, outputs, and components. From such ternary relations, which will by now appear very familiar to us, we shall extract a pair of *binary* relations, or oriented graphs, which interlock in a particular way. It is this pair of interlocking graphs that constitutes the most general relational description of the associated machine.

Our main result is concerned with the kinds of interlocking graphs, or relational descriptions, that can arise from machines in this way. This result can be elegantly phrased as follows. It is possible to define a generalized notion of *paths* through these interlocked graphs. Our result is: *in a relational description of a machine, there can be no closed paths.* As we shall see, it is this result that dooms the Cartesian metaphor, and much more than this, that exposes the inadequacy of contemporary physics itself as a vehicle for encompassing biology, and hence, material nature.

These are dramatic assertions, even more so in their various implications. Discussion of some of these will in fact constitute the remainder of this book. Nevertheless, they follow entirely from Natural Law (i.e., from the concept of a model) and the special nature of machines and mechanisms. Nothing else has been used in my arguments.

My conclusion is not, however, entirely negative. On the contrary, by virtue of the manner in which mechanisms fail to encompass organisms, more suitable kind of encodings automatically suggest themselves. These, as will be seen, involve *limiting processes,* which take us in a natural way out of the world of mechanisms or simple systems, into a larger world of *complex systems.* I shall discuss these issues in due course. For the present, however, I describe the relational properties of machines and their properties.

9B. The Basic Ideas

We now turn our attention to what we have called *machines;* natural systems such that (1) all models are simulable, and (2) there is a model of the system that is a mathematical machine.

Since a machine is a fortiori a mechanism, we have seen that it must have a unique largest model and a family of smallest models. The largest model, which we recall contains everything we can know about the machine, is the direct sum of the smallest ones. Hence we can introduce, once and for all, a set of states for the machine as a natural system, as we saw above. This much is true for any mechanism.

I will now invoke the further hypothesis that, in the set of all models, there is one that is a mathematical machine. Indeed, in what follows, I will use only one feature of this hypothesis, namely, *that we can make a distinction between software and hardware.* We will see what this distinction means, when reflected in terms of the states of the largest model.

Let us begin with a few words of heuristics. Intuitively, we think of the hardware of a mathematical machine as a *processor* of software. As such, this hardware has an inherent *polarity;* it has an input or afferent side, and an output, or efferent side. This polarity also reflects, and manifests, the flow of time in the system dynamics; the flow is *from* input *to* output, from afferent to efferent. With respect to the hardware, input and output are alike software, but the outputs reflect the past, and the inputs represent the future.

We also observe that, in material terms, hardware is *open* relative to software, and, in general, conversely. When taken together, they collectively constitute a single phase, in the Newtonian sense (see section 4H above). The dynamics of the machine is simply the recursive specification of next such phase in terms of present phase, as exemplified in the notion of *algorithm.* From the trajectories of phase so generated, we can extract or chronicle the transition from input to output, which we identify with the *operation* of the machine. In a sense, then, hardware constitutes tangent vectors to software, and conversely.

So we have the following: first, a categorical distinction between hardware and software; this arises simply from the fact that we are dealing with a machine. Then, a further distinction, imposed by the hardware on the software, a distinction between input (what the hardware *is going to* process) and output (what has already been processed). This last distinction is, however, neither categorical nor exhaustive; intuitively, insofar as "software" connotes anything about our machine that is not hardware, there can be plenty of software that is neither input nor output.

We now want to reflect these features of machines concretely on the set of states of the largest model, which in turn is the direct sum of the smallest models. First, the distinction between hardware and software, then the flow from input to output.

The distinction between hardware and software means that we can decompose the largest model into two big direct summands. Every state of the machine can thus be split into two blocks, one block pertaining to what

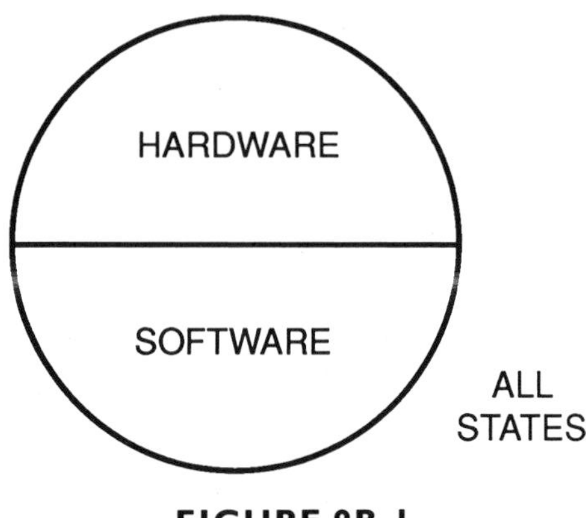

FIGURE 9B.1

we called hardware alone, the other pertaining to what we called software alone. Pictorially, we have figure 9B.1.

Every "particle" of the system belongs exclusively to one or the other of these summands. Thus, since by hypothesis we can write a state of the system in terms of the states of the constituent particulate models, we can partition such a state, say

$$(\ldots, x_{n-1}, x_n, x_{n+1}, \ldots) \qquad [9B.1]$$

into two blocks:

$$(\underbrace{\ldots, x_{n-1}, x_n}_{\text{HARDWARE}}, \underbrace{x_{n+1}, \ldots}_{\text{SOFTWARE}}) \qquad [9B.2]$$

This is, as we see, only a statement of fractionability; cf. section 8I above.

Next: As we noted, the hardware of the system (and more specifically, its polarity) imposes a further classification upon software, namely, a classification into *input, output,* and "everything else." Thus, we can pictorially refine the diagram of figure 9B.1 into figure 9B.2.

Correspondingly, in formal terms, we can partition a state of the largest model into

$$(\underbrace{\ldots, x_{n-1}, x_n}_{\text{HARDWARE}}, \underbrace{\underbrace{x_{n+1}, \ldots, x_r}_{\text{INPUTS}}, \underbrace{x_{r+1}, \ldots, x_s}_{\text{OUTPUTS}}, \underbrace{x_{s+1}, \ldots}_{\text{ALL OTHERS}})}_{\text{SOFTWARE}} \qquad [9B.3]$$

Roughly speaking, we have thereby reexpressed a state of the machine as a direct sum of four summands:

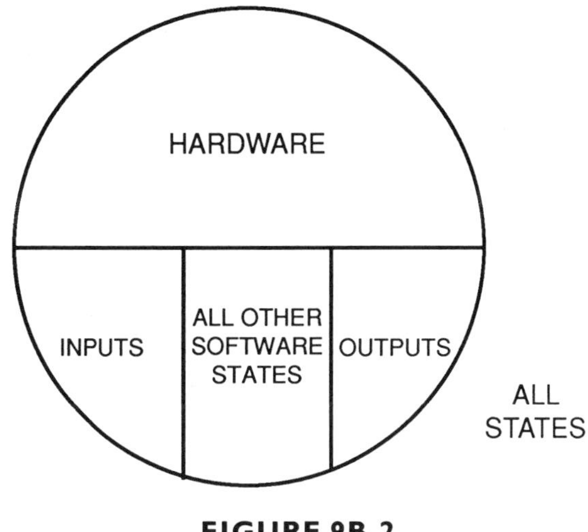

FIGURE 9B.2

1. its hardware
2. its inputs, the future environments of its hardware
3. its outputs, the pasts of its hardware
4. everything else, which we may identify with present phase of hardware + software

The last three items constitute software.

In terms of, say, a Turing machine, these summands correspond to:

1. the reading head of the machine
2. everything on the tape of the machine to the left of the reading head at an instant
3. everything on the tape of the machine to the right of the reading head at that instant
4. the scanned symbol at that instant

The whole machine is a direct sum of these four specifications. All we have done so far is to reflect these aspects of a mathematical machine back to the largest model of a natural system, and thence back to the natural system itself.

We now want to incorporate the dynamics of the machine, recalling the recursivity of state transitions in the machine as a whole. As we have seen, this can be expressed in terms of a flow imposed by the hardware on the software, from input to output. I will indicate these aspects diagrammatically by amending figure 9B.2 to figure 9B.3.

Here, the hollow-headed arrow denotes the actual software flow from input (afferent) to output (efferent); the solid-headed arrow denotes the

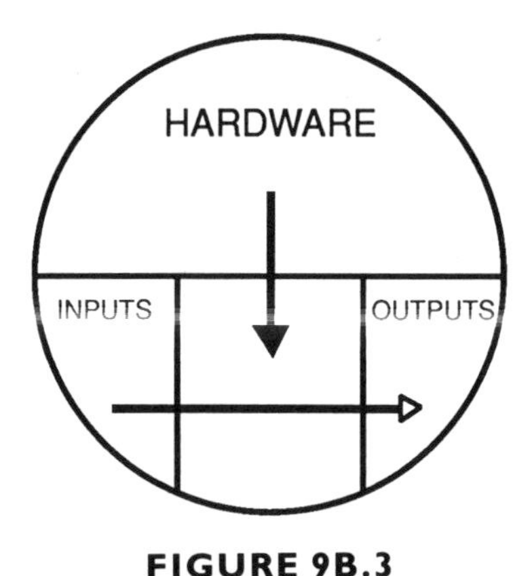

FIGURE 9B.3

induction or generation of this flow by the hardware. The solid-headed arrow thus expresses the relation between the hardware (which is just a set of states of something) and the software *flow* it induces, and it will play a central role in what follows. In terms of the state decomposition, the flow from inputs to outputs takes a particular form. We have seen that, in general, if a mechanism is a machine, then its states can be segregated into blocks or direct summands, i.e., put into the form

$$\text{state} = ((\text{hardware}), (\text{inputs}), (\text{other}), (\text{outputs})).$$

Since these blocks are direct summands, they are unlinked and can be selected ad lib. Let us take the particular choice

$$\text{state} = ((\text{hardware}), (\text{input}), 0, 0).$$

We can further choose (input) to be the most elementary possible, e.g., corresponding to a situation in which only one symbol appears on the input tape to a Turing machine. Then, in general, the software flow imposed by hardware on software will take us recursively through states of the form

$$(\text{hardware}, (0), (\text{other}), (0))$$

to an ultimate state of the form

$$(\text{hardware}, (0), (0), (\text{output})).$$

We thus see the germ of a functional relation between such outputs and their corresponding inputs. This will provide us in a moment with the essence of the *relational description* (or at any rate, with one relational description) of the machine itself.

Let us then digress for a moment, and look again at what it means formally to express a mapping relation. Let us look then at the familiar diagram

$$f : A \to B \qquad [9B.4]$$

or

$$A \xrightarrow{f} B$$

Let us note the evident fact that there are four elements or ingredients in the diagram, four symbols that make it up. Explicitly, we have

$$\begin{aligned} A &= \text{domain of the mapping,} \\ B &= \text{range of the mapping,} \\ f &= \text{the name of the mapping.} \end{aligned}$$

and

$$\longrightarrow \; = \text{the name of a } \textit{ternary relation} \text{ between } f, A, \text{ and } B$$

We usually *identify* "f" with "$\longrightarrow$", so that the four distinct symbols in the mapping diagram [9B.4] really involve only three corresponding referents. We do this because we identify f itself with a *binary* relation between domain A and range B. These identifications are so routine that no one ever notices them; indeed, in ordinary circumstances they are entirely innocent and devoid of consequence. But it is *abus de langage,* nevertheless, and will become of great moment for us now; we are about to confront a situation in which these conventional identifications no longer apply, one in which each of the four symbols in [9B.4] will have *an independent referent.*

In fact, we are going to use the expression

$$A \xrightarrow{f} B \qquad [9B.5]$$

as an abbreviation for

$$\begin{array}{c} f \\ \big\downarrow \\ A \longrightarrow\!\!\triangleright\; B, \end{array} \qquad [9B.6]$$

which, as the reader will note, is simply figure 9B.3 with the inessentials erased. That is, we now mandate:

$$\begin{aligned} A &= \text{name of set of } \textit{inputs,} \\ B &= \text{name of set of } \textit{outputs,} \\ f &= \text{name of } \textit{hardware,} \end{aligned}$$

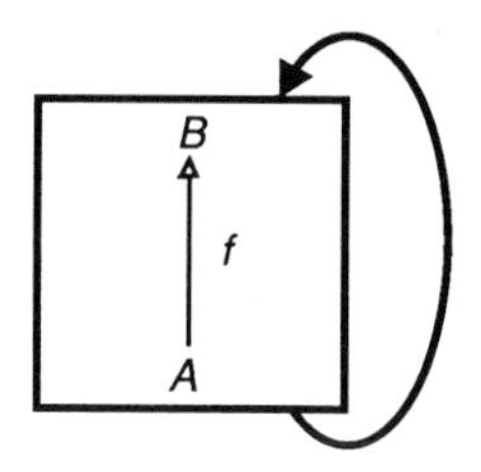

FIGURE 9B.4

and

$\longrightarrow$ = name of *flow* on software, from inputs to outputs, induced by hardware *f*.

Parenthetically, we may note that the traditional ignoring of the fact that a mapping diagram like [9B.5] is an abbreviation, which can identify (and hence equivocate between) incomparable situations, has been responsible for a certain amount of mischief in mathematics itself. Thus, whereas Gertrude Stein could claim that "rose is a rose is a rose," it is not always true that "mapping is a mapping is a mapping."

In terms of direct sums of states, we have the translation of [9B.3] above into the diagram [9B.5] or [9B.6] as follows:

$$(\ldots, x_{n-1}, x_n, x_{n+1}, \ldots, x_r, x_{r+1}, \ldots, x_s, x_{s+1}, \ldots)$$

HARDWARE	INPUT	OTHER	OUTPUT
=	=	=	=
f	A	$\rightarrow$	B

It cannot be stressed too strongly that, in these considerations, the hardware *f* and the flows it induces on software are fundamentally different things; they encode entirely different aspects of the natural system they model. I have tried to make this clear by exhibiting the encoding itself in various ways; I will try one last time by rewriting [9B.6] in terms of the modeling relation itself; see section 3H above. In terms of that discussion we should think of the right-hand box (i.e., the model) as being of the form shown in figure 9B.4. That is, all of the *states* (i.e., hardware plus software) go *inside* the box, as does the flow from input to output. The *generation* of that flow by the hardware (i.e., the black arrow) is what sits outside the box as the inferential structure, as indicated. I discuss some of the causal correlates of this picture in subsequent sections.

In any case, we have passed with some difficulty from a natural system

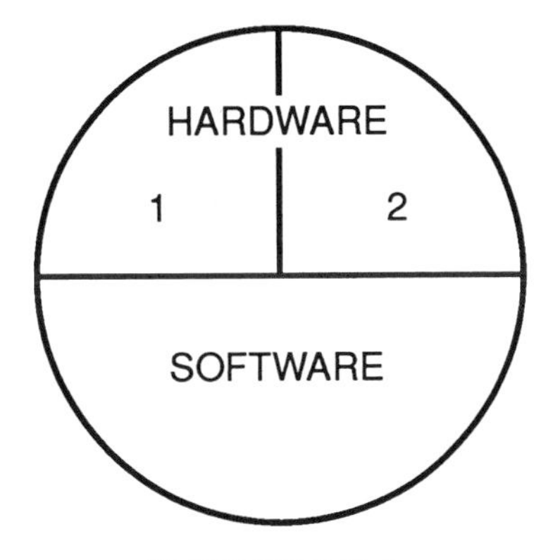

FIGURE 9C.1

that is a machine to a formal representation of it, of the form [9B.6], or, in abbreviated form, [9B.5]. This is what I shall call a *relational model* of the machine. As we see, there does not seem to be much left of the machine itself in this version of it. For instance, we see no explicit encoding of time, have no dynamics in the diagram. The diagram does, however, embody the basic *polarity* of the machine, the progression in time from afferent to efferent, from input to output. This will turn out to be the essential temporal feature for us, not time divided into minutes and seconds, but time *encoded as a chase through a diagram.*

Note also that our discussion so far has been at the most elementary level, to fix ideas. We have culminated our discussion with the simplest kind of *diagram,* and one of the features of such diagrams is that they are *open-ended;* domains in a diagram can be ranges of other things (other mappings); ranges can be domains; mappings can be factored; etc. Most important of all, the diagram can be separated from the specific material structure that gave rise to it; this is in fact the essence of relational description. The diagram survives, even when nothing remains of the interpretations with which we have associated them above. From these elementary facts, entirely new science will spring. But we shall proceed a step at a time. Having taken the first step, let us now take the next.

9C. The Second Step

In the considerations so far, we have relied on an absolute but gross partition of the states of a mechanism into hardware and software. We will now see what happens when we can refine these distinctions somewhat.

So let us look at the diagram (figure 9C.1). The states we previously

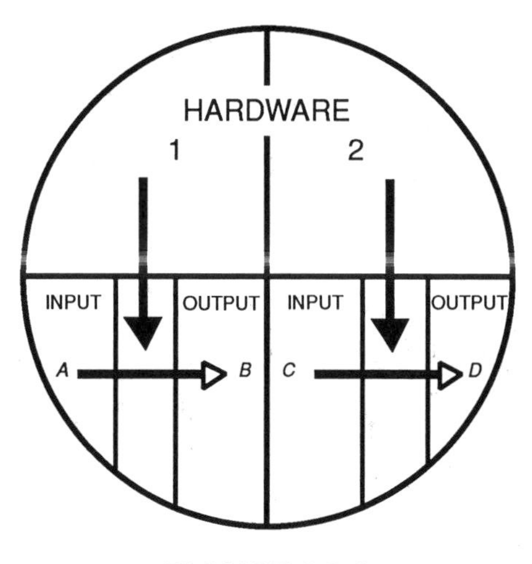

FIGURE 9C.2

designated as "hardware" have now been partitioned into two subblocks. We want to endow each of these with the properties previously enjoyed by the whole block; i.e., we want to be able to call each subblock "hardware" in its own right.

Well, what does hardware do? As we have seen, the essence of it is that it generates flows from inputs to outputs in software. Thus, a partition of hardware into two pieces, each of which is itself hardware, involves a further partition of *software*. Specifically, each piece of hardware must be given its own share of software, its own inputs and outputs, and its own flow between them.

Let us look at several possibilities. The simplest is diagrammed in figure 9C.2. Here I have correspondingly partitioned the software states into two large blocks, each corresponding to a hardware block; i.e., I have endowed each hardware block with its own software. Then, within each of these software blocks, I have further partitioned into inputs, outputs, and others. The result is, in effect, two copies of figure 9B.3 above, placed side by side. Each of them separately give rise to a relational diagram:

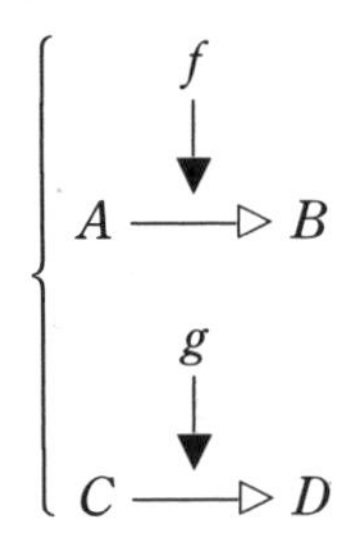

in the fashion we have indicated above, or in the more customary abbreviated form, simply

$$\begin{cases} A \xrightarrow{f} B \\ \\ C \xrightarrow{g} D \end{cases}$$

Somewhat more interesting is the case in which B and C coincide. That is, the *outputs* of the first block are also the *inputs* of the second. In this case, we evidently have the diagram

$$\begin{array}{ccccc} & f & & g & \\ & \downarrow & & \downarrow & \\ A & \longrightarrow\!\!\triangleright & B & \longrightarrow\!\!\triangleright & C \end{array}$$

which abbreviates to the simple *composition* of the two mappings f, g:

$$A \xrightarrow{f} B \xrightarrow{g} C.$$

This is the first nontrivial (though still very simple) relational model we have encountered; see section 5G above. Despite its simplicity, it already embodies some features that are worth calling attention to. The first thing is that it is a diagram of mappings, which embodies in its *form* the central features of the material system (in this case, the machine) with which it is associated. Second: once given the form of the diagram, it is clear that *chasing an element of* $a \in A$ *through the diagram* encodes the essential features of system dynamics. Third: it is clear that although we began our discussion with a *specific* machine, a specific natural system with its own set of states, the *same* relational description can arise from many different natural systems, systems that may otherwise look quite incomparable.

The situation we have just discussed was obtained from figure 9C.2 by identifying the sets of software states labeled B and C in that figure. We could make other identifications as well. For instance, we can identify A and C. The reader can easily show that such an identification gives rise to a completely different relational diagram, of the form

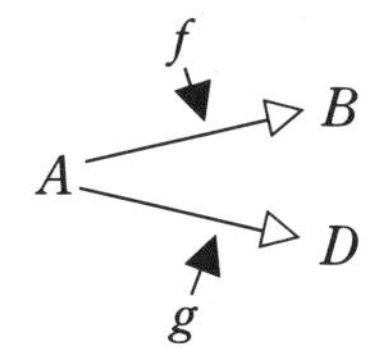

or, as customarily abbreviated

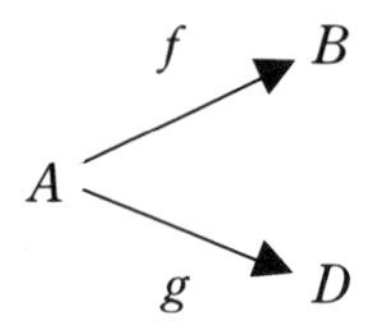

Let us see what these decompositions mean in terms of *states.* As we recall, we started from the fact that, since we are dealing with a mechanism, we can fix a set of states of the form

$$(\ldots, x_{n-1}, x_n, x_{n+1}, \ldots)$$

once and for all. From the fact that we are dealing specifically with a machine, we can fraction these states uniquely into hardware and software:

$$\underbrace{(\ldots, x_{n-1}, x_n,}_{\text{HARDWARE}} \underbrace{x_{n+1}, \ldots)}_{\text{SOFTWARE}}$$

and then further partition the software as we indicated.

We are now supposing that we can further partition the *hardware* into two summands; i.e., we can write

$$\underbrace{\underbrace{(\ldots, x_r,}_{\textcircled{1}} \underbrace{x_{r+1}, \ldots, x_n,}_{\textcircled{2}}}_{\text{HARDWARE}} \underbrace{x_{n+1}, \ldots)}_{\text{SOFTWARE}}$$

Along with this comes a further partition of the software, so that each of the hardware blocks has its own inputs, outputs, etc. It is clear that, at this level, the relational description is simply an abbreviation for, and a representation of, these partitions.

To have a name for such summands of hardware, possessing the properties we have indicated, we will henceforth call them *components* (see section 5D. above), here specifically tied to states.

The language of states and state partitions becomes increasingly cumbersome and inappropriate as the number of such components increases. I believe the reader can already see, in the case of two components we have been discussing, how unwieldy it is. The relational description is always, however, at our disposal and can (at least under present circumstances) always in principle be reexpressed as a partition of states into hardware and software, of hardware into components, and software into associated inputs, outputs, etc. In what follows, I shall use the relational description exclusively, without further comment, and (significantly) without reference to the system of states it represents.

There is one further situation it is convenient to mention here. It first arises in a situation where we have three components, i.e., three mappings

$$f \ : A \rightarrow B,$$
$$g \ : C \rightarrow D,$$
$$h \ : E \rightarrow F.$$

I remind the reader one last time that the mappings f, g, h are abbreviations for *hardware;* the sets $A, \ldots, F$ represent *software,* and the arrows abbreviate flows from input to output. Just as in the two-component case, a variety of relational diagrams is possible, depending on what relations exist between the software sets $A, \ldots, F$, i.e., on how inputs of one component are related to outputs of others.

One such relation that is now possible is the following:

$$E = B + D.$$

That is: the input set to one component (in this case, the one represented by the mapping h) is the direct sum of the outputs of the other two. It will become important to have a notation to express such a situation.

Formally, of course, we could express it by means of a diagram of mappings of the form

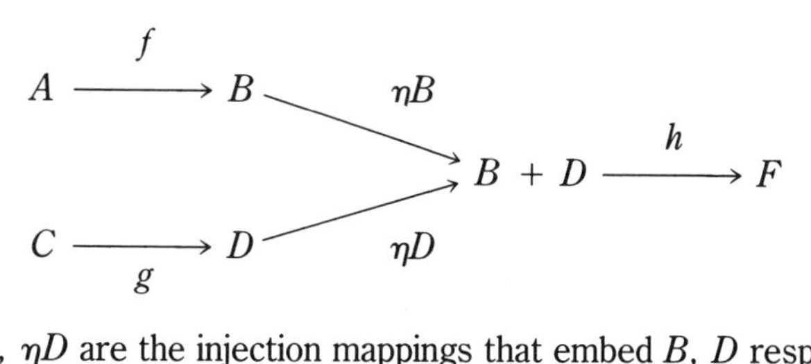

where ηB, ηD are the injection mappings that embed B, D respectively in their direct sum. However, this would require introducing into the diagram two mappings (ηB and ηD) that do not decode into components. It is awkward to introduce two *kinds* of mappings, mappings with distinct status, into these relational descriptions. Accordingly, I adopt another notational device. In a situation such as that described above, I write instead

$$\left.\begin{array}{l} A \xrightarrow{f} B \\ \\ C \xrightarrow[g]{} D \end{array}\right\} \xrightarrow{h} F$$

and use the bracket to indicate that the domain of the mapping h is the direct sum of the bracketed sets.

We can obviously extend the discussion to any number of components (hardware), governing any kind of input-output software flows. With each such situation, we can associate a diagram of mappings, the corresponding relational model. Henceforth, I shall call such a diagram of mappings, such a relational model, an *abstract block diagram* of the machine with which it is associated.

9D. Entailment in Machine Models

Let us recall that the whole purpose of modeling is to bring causal entailment, in the external world of natural systems, into congruence with inferential entailment in a formal system or model. Thus, to assert that a natural system has a model that is a mathematical machine is to say something very special about the operation of causality in that natural system (or at least, about causality in the part of the natural system that encodes into the mathematical machine model). This in turn is embodied in the *form* of the relational model, the abstract block diagram, which we have constructed in the preceding section. It will turn out that the essential features of what we say now generalizes far beyond machines, in fact, to general mechanisms. But it is already visible here, and it is convenient to introduce the basic ideas in this special context.

Let us begin with a single component

$$A \xrightarrow{f} B,$$

which, as we recall, is an abbreviation for a more detailed encoding of the form

$$\begin{array}{c} f \\ \downarrow \\ A \longrightarrow\!\!\triangleright\, B \end{array} \qquad [9D.1]$$

This diagram already embodies what we have taken as the essential feature of a mathematical machine, namely, the absolute partition between *hardware* and *software.* This has translated in the diagram into the distinction between the mapping f (which, as we recall, is a name for the states we assign to hardware) and the sets A, B (the domain and range of f respectively), which designate certain sets of software states. The states so named are disjoint summands.

I have already discussed causality, in the context of general mechanisms, at great length. As we have seen, it is the essence of a mechanism that causality manifests itself therein entirely through the recursiveness of its state transition sequences, i.e., in the entailment of next state by present state. Thus, if

$$s(t) = (\ldots, x_{n-1}(t), x_n(t), x_{n+1}(t), \ldots)$$

encodes the state of a mechanism at instant t, and we ask the question "why $s(t)$?" we can meaningfully answer it by saying "because $s(t_o)$ initially." But in general, we cannot meaningfully ask a question like "why $x_n(t)$?" in this kind of context. Such questions acquire causal status in mechanisms only in terms of the states of the mechanisms as a whole.

The situation is different if we are dealing with a machine. In this case, we superimpose on a general mechanism the partition into hardware and software, which as we have seen, reflects itself in a corresponding partition of mechanism *states* into separate blocks. We saw above that, in a machine, the recursive state transition sequence we always have in a mechanism manifests itself in terms of flows *between* these blocks. Thus, in a machine, the notions of causality we inherit from general mechanisms can be recast, in terms of what goes on *between these blocks,* a possibility that is not generally available in mechanisms per se. We will now see what this means.

As always, causality involves dealing with the question "why X?" In the present situation, the natural question to ask is "why $f(a)$?" i.e., to regard an element $f(a)$ in B as *effect* and inquire into its causes, into what entails it. There are obviously two distinct but correct answers we can give to this question: namely, *because* $a \in A$, and *because* f. As we saw in section 5H above, the first answer is associated with *material causality,* the second with *efficient causality*. In terms of the notations we have been using, we can (see [5H.1] above) sum up both of these in a single expression:

$$f \Longrightarrow (a \Rightarrow f(a)). \qquad [9D.2]$$

which is, of course, just another way of rewriting the expression [9D.1]. In this form, however, we exhibit in the clearest manner possible that, in mathematical machines, efficient causality and material causality are *segregated* into disjoint structures; specifically, hardware is the embodiment of efficient cause, while material cause is embodied in software. (Later, when we bring in the idea of *program,* which is also embodied in software, we will also see that the category of *formal cause* is likewise segregated). It will be noted that, at root, all this arises entirely from fractionability, which is itself a corollary of mechanism.

Let us consider a slightly more complicated situation, namely, the diagram

$$A \xrightarrow{f} B \xrightarrow{g} C$$

which abbreviates

$$\begin{array}{ccccc} & f & & g & \\ & \downarrow & & \downarrow & \\ A & \longrightarrow\!\rhd & B & \longrightarrow\!\rhd & C \end{array} \qquad [9D.3]$$

Now suppose we ask "why $gf(a)$?" where $gf(a)$ is an element of C. In the notation we have just used, we can succinctly answer this question; $gf(a)$ because

$$g \Rightarrow\!\!\blacktriangleright (f(a) \Rightarrow\!\!\rhd gf(a)). \qquad [9D.4]$$

Furthermore, we already have [9D.2] above. The two can be combined into a single (rather cumbersome) expression, of the form

$$g \Rightarrow\!\!\blacktriangleright [\{f \Rightarrow\!\!\blacktriangleright (a \Rightarrow\!\!\rhd f(a))\} \Rightarrow\!\!\rhd gf(a)] \qquad [9D.5]$$

which again is just a rewriting of [9D.3]. We can, however, note that the segregation of causal categories still holds (the solid-head and the hollow-head arrows clump together, the former on the left, the latter on the right).

The ideas I have exemplified above are clearly perfectly general, applicable to any machine. To repeat: once we say that a natural system possesses a model that is a mathematical machine, we have placed what will turn out to be devastating limitations on the operation of causality in that system (or at least, on what is encoded into the machine model). Specifically, we have segregated the causal categories into disjoint structures.

Since this will turn out to be the Achilles' heel of machines, and indeed of general mechanisms, I recast it in yet another form, a graph-theoretic form. To fix ideas, let us slightly modify the diagram of [9D.3] above, so that we have a conventional oriented graph of vertices and arrows (figure 9D.1). We can do this because the vertices (f, g, A, B, C) are names for *disjoint* sets of states of our machine. The main feature here is *the absence of certain kinds of paths* in this graph. Specifically:

If a vertex in this graph originates a solid-head arrow, it cannot terminate a hollow-head arrow.

All this can be summed up succinctly in the following form: the solid-head arrows, and the inferential rules they represent, are themselves

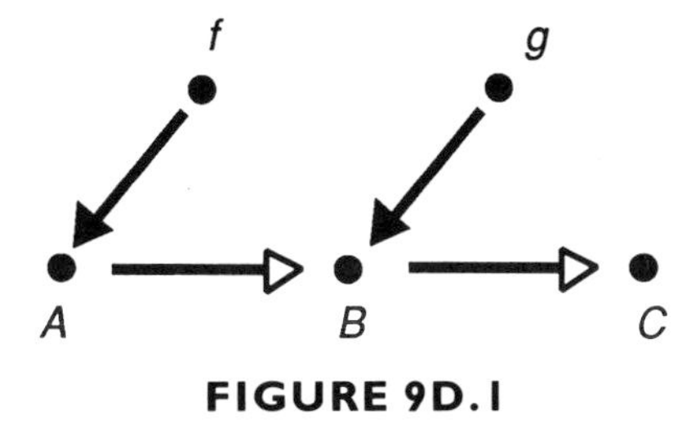

FIGURE 9D.1

unentailed in the machine. That is: we cannot answer a question like "why *f*?" within the system, except by means of *the category of final causation* ("*f* because *f* entails . . .").

The empoverishment of causality in machines is their most important and most characteristic feature. I shall return to it again and again as I proceed.

9E. The Third Step

In what follows, we will in effect begin to ask the forbidden question "why *f*?" We have seen that such a question has no answer, within the confines of the machine model within which *f* appears. But in a wider context, why should we not ask this question? After all, *f* is only the name we have assigned to a set of states (hardware). Even in that context, *some* states are entailed (namely, the software states we called outputs). Let us recall that we have nowhere assumed that the machine model we have been using is already the largest model M^{max}; i.e., encodes *every* state of the system it models. Maybe it is only a direct summand. Maybe we can enlarge it; maybe we can find a bigger model, with more states, in which the question "why *f*?" has an answer. And maybe this bigger model is itself a mathematical machine.

In the context of relational models, which are just arrays of sets and mappings, the question is easy to address, if we forget the fact that the sets and mappings in them are names for sets of states. Once we do that, it is child's play to identify the *elements* of a set A with *mappings* $f : B \rightarrow C$. Thus, any mapping of the form

$$\Phi : X \rightarrow A$$

will have *outputs* that are themselves mappings of the form

$$f : B \rightarrow C.$$

We have thus *entailed f;*

$$\Phi \Rightarrow (x \Rightarrow \Phi(x) = f). \qquad [9E.1]$$

Of course, now Φ is unentailed. But if we can entail Φ, we can entail f; maybe, if we are clever, we can arrange matters in such a way that *every* mapping in a suitably enlarged system is entailed by other mappings already in the system. In this fashion, we glimpse a nice, tidy situation in which everything in the system (or almost everything, at any rate) is already entailed by something else in the system.

All this is entirely meaningful, from the standpoint of *relational* considerations. What we shall now show is that this is precisely what *cannot* happen within the confines of machines, or even of general mechanisms. From this, we will conclude that, if it does happen, then what makes it happen *cannot be a machine, or even a mechanism.* And finally, of course, we shall argue that it does happen. It happens in biology.

In causal terms, [9E.1] above allows us to answer the question "why f?" with "because Φ"; as we have seen, this kind of answer involves the category of *efficient causation:* ϕ is an efficient cause of the effect f. We can (literally) illuminate [9E.1] by writing it as we did [9D.2] above:

$$\Phi \Rightarrow (x \Rightarrow \Phi(x) = f) \quad [9E.2]$$

and thence, in graphical terms, as

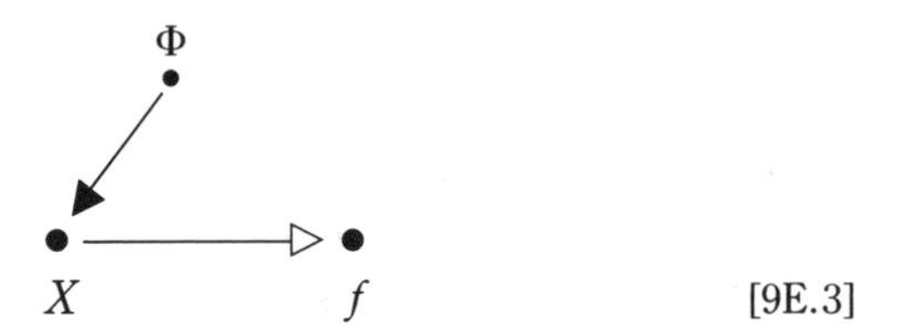

[9E.3]

Now we already have the graph

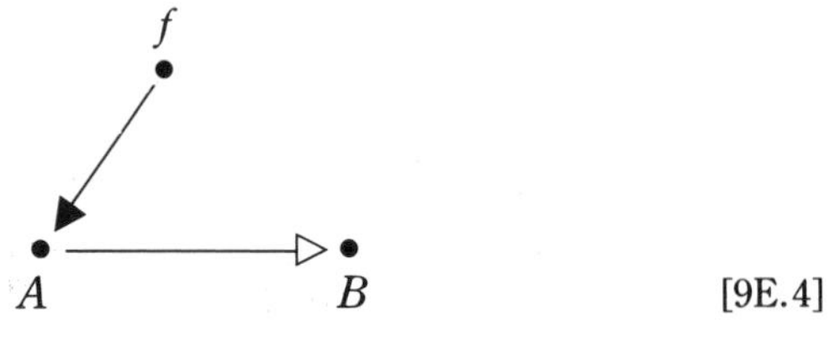

[9E.4]

and there is nothing to prevent us from putting these two graphs together to obtain figure 9E.1.

Let us contemplate this graph for a moment and inquire into what it connotes. First, it is obviously built out of two subgraphs, one that is the relational model of a machine, and another that looks formally just like it. The only difference between them is in *interpretation,* in decoding; in particular, what was hardware for the former has become software for the latter.

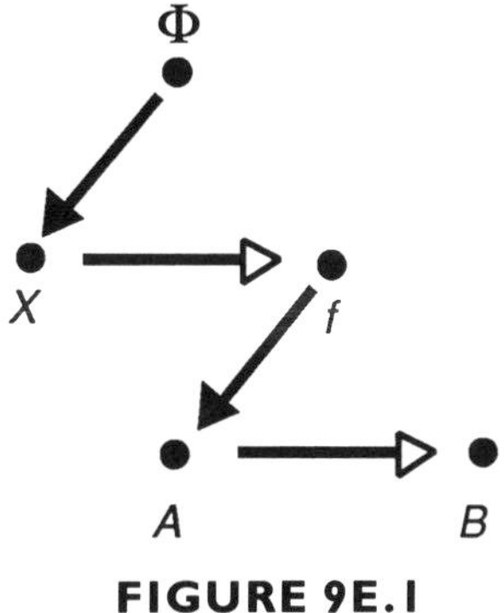

FIGURE 9E.1

I am going to take the position that the graph displayed in figure 9E.1 is itself a (relational) model of *something,* i.e., of some natural system, and interpret it accordingly, using the ideas I have already developed.

Evidently, the vertices in this graph must, in a mechanism, be names for *disjoint sets of states.* The open head arrows denote flows between these sets, as we have seen, and connote a category of material causation from inputs to outputs. The solid head arrows connote efficient causation of outputs (intuitively, the "processing" of software by hardware). In particular, it is clear that a natural system represented by figure 9E.1 in which we can answer the question "why f?" *must have more states* that the system represented by [9E.4] alone. Namely, it must have a set of states we can assign to Φ itself, and in general, others corresponding to the software of Φ. These must, by virtue of everything I have already said, be direct summands, which we tack on to the states already named by f, A, and B.

The next thing to note is that, *as a formalism* the graph in figure 9E.1 above can itself be regarded as a mathematical machine, of the form

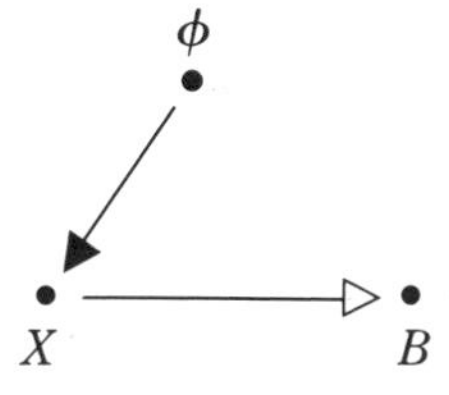

if we *interpret* the composite

[9E.5]

as a single software flow, with f as *program.* But clearly, such an interpretation would profoundly obscure the causal features of the original diagram; the reader can see this by asking the questions "why f?" and "why $f(a)$?" in the two situations. This clearly arises because the hardware/software distinctions implicit in [9E.5] are not compatible (indeed, are *inconsistent* with) those arising from [9E.4] and figure 9E.1.

Indeed, under this interpretation, [9E.5] describes a *simulation* of f by Φ, rather than an entailment of f by Φ. Nothing could more starkly illustrate the anomalous character of simulation and the mischief that can arise through its confusion of causal categories. I remark parenthetically that the confounding of simulation (computation) with *construction,* which lies at the heart of, e.g., von Neumann's well-known discussion of "self-reproducing automata," arises precisely here and rests entirely on the equivocal and inconsistent hardware/software distinctions to which I have just called attention.

The distinction between hardware and software, then, is a completely meaningful and legitimate one, but it must be maintained *consistently* when decoded back to a natural system. What I have concluded, then, is that it cannot be maintained consistently between the two diagrams [9E.4] and figure 9E.1. Hence, if the former is a mathematical machine, the other is not, though it is still a perfectly good *mechanism.*

Let us now return to the main line of the argument. I have in effect shown that, given a machine whose relational model is [9E.4], we can find a larger system (generally, only a mechanism) in which f is itself entailed, and which has a relational model of the form shown in figure 9E.1. The new system is larger in the sense that it must, as a mechanism, have *more states* than we had before; these comprise direct summands to be added to those represented in [9E.4] itself.

This argument can clearly be extended to an arbitrary block diagram, built of *components* that are each of the form [9E.4]. If we do so, we conclude that, given any abstract block diagram, which comprises a relational model of a machine, we can find a larger system in which any component g can itself be entailed. That is, in the larger system, we can answer the question "why g?" with an answer of the form "because Φ_g", and this answer represents the category of *efficient causation* for g as effect.

I am now going to iterate these ideas. Specifically: in the system we have just constructed, we can answer all the questions "why g?" all right, but now we have a family of questions, of the form "why Φ_g?" that can be entertained. Once again, these are generally unanswerable within the system; the ϕ_g are generally *unentailed* in the new system, just as the original components g were, in the original system. In other words, we have answered our first set of questions, but at the cost of posing new ones.

I will now show that the concept of a mechanism is inconsistent with these iterations.

9F. The Central Argument: The Limitations of Entailment in Machines and Mechanisms

Before embarking on the details, let me pause to exhibit pictorially what I am doing. We have been considering certain natural systems, which we suppose to be mechanisms (every model simulable), and further, to be machines (at least one *model* is a mathematical machine). We then used the hardware/software distinction to obtain a relational description of such a machine, a graph of sets and mappings. In this context, the sets and mappings were abbreviations for certain disjoint sets of states (direct summands). In doing this, we have made heavy use of all of our hypotheses regarding both machines and mechanisms.

As we noted, the entailment structure in such a machine did not allow us to pose the question "why *f*?" for any mapping in the abstract block diagram of a machine. In particular, we cannot answer this question in terms of a category of efficient causation. Yet the question is formally meaningful. To answer it means appending more mappings to the original abstract block diagram, mappings whose ranges are themselves sets of mappings.

We thus obtain a new, larger graph of sets and mappings, which looks like the abstract block diagram of *something.* That something is a larger natural system, if it is anything at all. If there is no natural system to which this larger relational model corresponds (i.e., of which it is a model), then we must conclude that the hardware of our original system admits no category of efficient causation, i.e., that this hardware cannot be made or fabricated or entailed by anything else in the external world. We cannot simply discount this possibility out of hand, at least not yet; but it is clearly unpleasant. Thus, we will simply set it aside for the time being; in effect, we shall suppose (although Natural Law, because it goes *from* ambience *to* formalism, does not quite entitle us to do this) that any component in a machine, anything that can be relationally encoded as a mapping *f* in the manner we have indicated, can itself be entailed by something in the external world. Moreover, that "something" can itself be regarded as a component, i.e., imaged relationally by a mapping Φ. The new component precisely answers the question "why *f*?" in the sense of efficient cause: because Φ.

The reader should note that this is the first time I have argued backward, from a formalism to the *existence* of something in the external world that is modeled by the formalism. In particular, I have argued from a machine to another natural system, in which its *components* are themselves

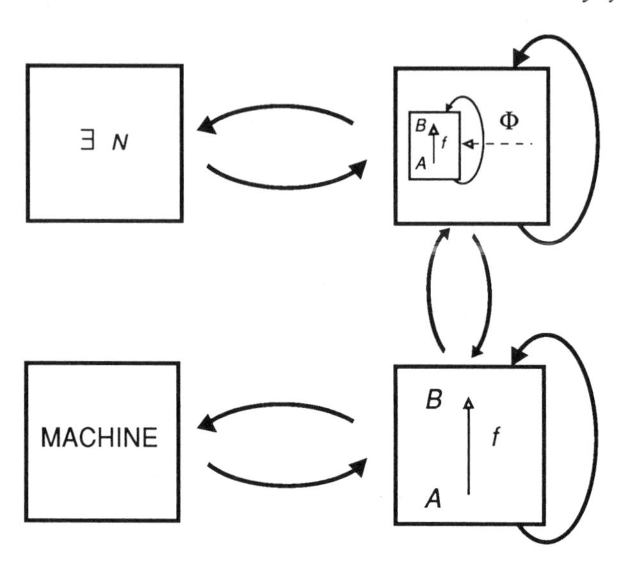

FIGURE 9F.1

entailed. I have thus implicitly invoked a new kind of entailment between natural systems in the external world, one we have not seen before. This is in fact a form of what is called in philosophy IMMANENT CAUSATION and involves ontological, as well as epistemological, considerations. We will not get to God this way, as Aristotle and others thought they did, but as we shall see, it will take us out of the class of mechanisms.

What we end up with, then, is a diagram of the form figure 9F.1, in which the upper left-hand box designates the larger natural system in which the original hardware is efficiently entailed. In fact, we can now throw away the original machine, since it is *included* (in the obvious sense) in the new system N. We thus have a new natural system N, with a number of separate (relational) models (one of which is a model of the original machine). We suppose that N too is a mechanism, and hence, in particular that the original machine is a *direct summand* of the larger system N.

Now we want to take the next step; we want to entail Φ according to a category of efficient causation. Formally, this is very easy; we simply want another mapping Φ that has Φ in its range. In simplest relational terms, we want to go from the diagram of figure 9E.1 above to a larger diagram of the form figure 9F.2.

The obvious thing to do is to embody this new mapping Φ in another disjoint direct summand of states, as we did before, and simply repeat the argument we have already used. This would give us a larger system N'

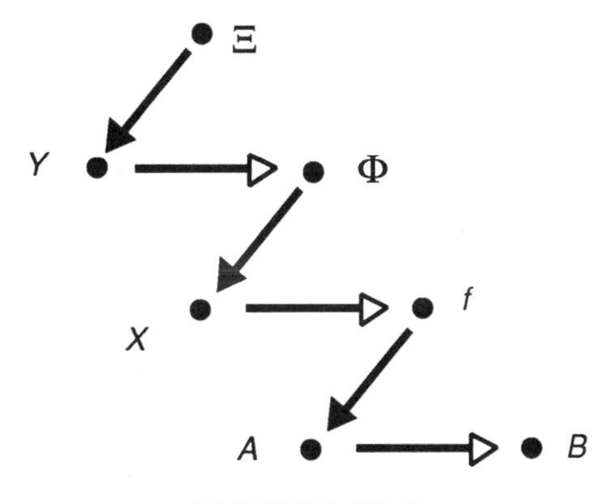

FIGURE 9F.2

(i.e., one with more states, more direct summands). In this larger system, we could answer both the questions "why *f*?" and "why Φ?" in terms of categories of efficient causation. But the question "why Φ?" would now be askable and unanswerable.

If we do this, we see the obvious infinite regress forming. We would obtain thereby an infinite sequence of mechanisms $N^{(n)}$, in each of which there would always be some unentailed components. Only in the *limit* $n \to \infty$ could we talk about *every* component being efficiently entailed, i.e., in which every question of the form "why Ψ?" would have an answer in the system. But this limit obviously *cannot be a mechanism;* a mechanism must have a largest (simulable) model, and the limit does not.

Thus, the situation we have been considering does indeed take us out of the class of mechanisms; we cannot keep adding direct summands at each step, to entail the components we had to add at the previous step, and stay within the class of mechanisms. But there are other, more subtle possibilities, and we must now consider these.

These possibilities arise as follows. Let us, for instance, compare once again the graphs shown in figure 9E.1 and figure 9F.2. In intuitive terms, we need states for Φ, and states for its domain *Y;* we recall that every vertex in such a graph was just a name for a set (indeed, for a direct summand) of states. In the argument above, we simply threw in more states, more direct summands, for these new vertices to name. But it may be that *we already have enough states,* i.e., that Φ is *already* efficiently entailed in the diagram [9E.1]. As we shall now see, this possibility means that we can entail Φ, not by *adding* more states as we did above, but rather by drastically *constraining* the ones we already have.

Note also (though we will not be using it much in this volume) that the

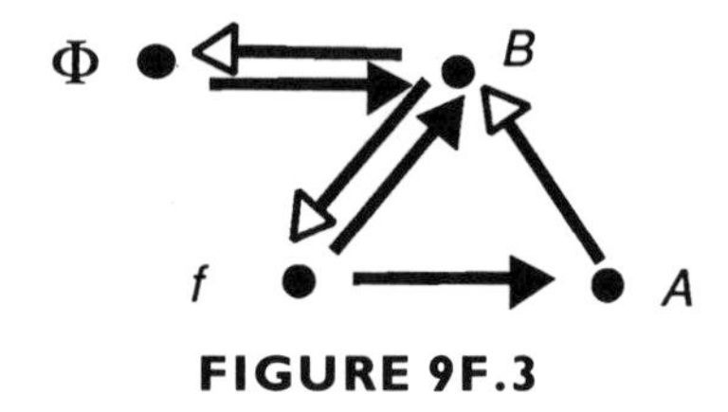

FIGURE 9F.3

partitioning of states I have described is closely related to the formal partitioning of software into data and program (see section 7D above). The ability to make such partitions is thus akin to *programmability* in a mathematical machine. In an important sense, I am saying that the more constrained a mechanism is, the more programmable, the more like an organism it appears. (The reader should recall the arguments of Elsasser; see section 1A above.) Note also that the idea of program is tied to the category of *formal causation.* These are indeed highly significant observations, but it would take us too far afield to pursue them in the present context.

These possibilities are easy to see in their purely formal counterparts. So far, we have been profligate in throwing vertices into our graphs and assigning them different symbols, different names. Thus, in figure 9F.2, we have the symbols A, B. X, Y, etc. But purely formally, why could we not mandate, e.g., $X = B$? or $Y = X = B$? Why could we not even mandate $f = \Phi$? If we make all these identifications, we can correspondingly make the graph of figure 9F.2 collapse to something of the form figure 9F.3. In such a diagram, *everything* (except A, the original inputs) is efficiently entailed, without the need for adding more states, merely by imposing or mandating additional restrictions (constraints) on those we already have.

Intuitively, the imposition of constraints of this type means that we are, in effect, going to *smaller models* (fewer states) rather than to bigger ones as we did before. "Smaller" here means that, instead of considering arbitrary states of N, the natural system represented in figure 9E.1, we consider only the ones satisfying the additional constraints leading to the larger diagram shown in figure 9F.3.

There is clearly a lot going on in this diagram, as compared with the more leisurely one (figure 9F.2) from which it was obtained. In particular, the component f is now rather busy, in terms of efficient causation; it answers *two* questions: "why b?" and "why Φ?" (B is itself correspondingly busy with regard to the category of material causation, but this is not of present concern.) Thus, in relational terms, *f has two functions* in the diagram: (1) to make $f(a)$ from a $\in A$, and (2) to make $\Phi = f(b)$ from $b \in B$. In terms of the original natural system N of which figure 9F.3 is a

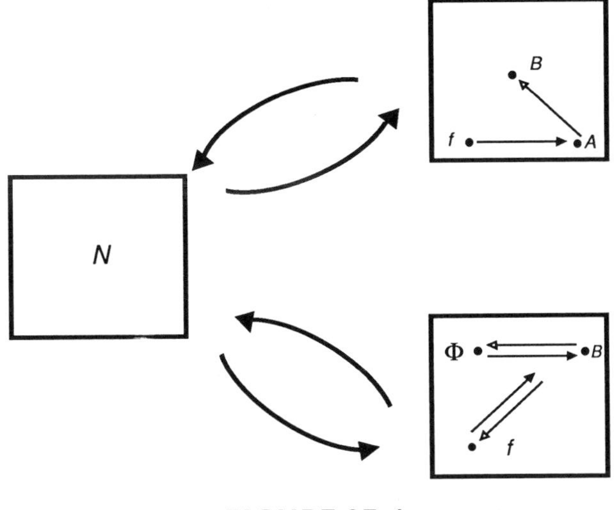

FIGURE 9F.4

model, the symbol *f,* and the set of hardware states it represents, are being encoded in *two different ways.* One of them is the original subgraph

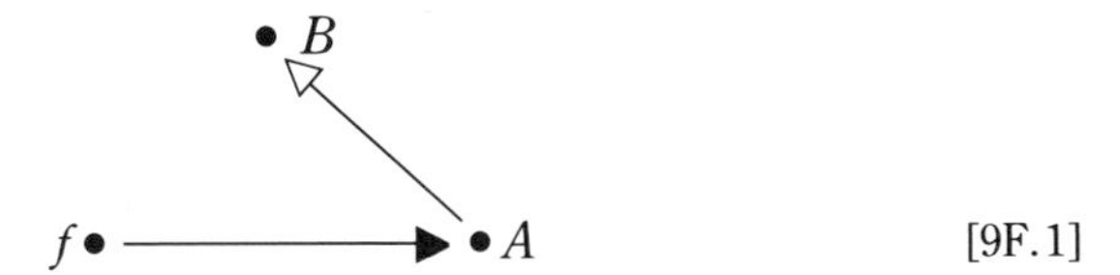

[9F.1]

and the other is represented in the subgraph

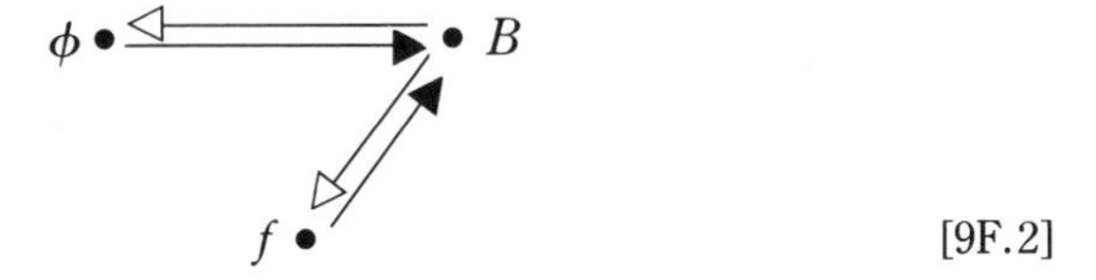

[9F.2]

We can already see that these diagrams are peculiar, but so far, they are completely consistent with the hypotheses of machine and mechanism. Recall that *f* is the name we have assigned to a set of states (hardware); as the origin for a solid head arrow, it means that these hardware states are to be identified with flows (i.e., with tangent vectors) on other states. It is precisely the fact that we are now forcing *f* to induce more than one such flow that embodies the *constraints* we have imposed, in going from figure 9F.2 to figure 9F.3. In these terms, then, we are encoding the same set of hardware states into (1) "tangent vectors" on the input states *A,* and (2) "tangent vectors" on the output states *B*.

Now neither of these encodings is visible to the other. We have in effect the situation shown in figure 9F.4, in which our natural system N is encoded into two different models; in each of them, f sits as the origin of a solid head arrow (i.e., answers a question "why?" in terms of a category of efficient causation), but these are two different arrows, in two different models. The question now is: what do these two different encodings of the same set of states decode back into? What do these separate encodings, into two different *functions*, connote about the set of states, which f also names?

In a nutshell, we must have that, *as a set of states of N, f itself splits into two direct summands,* one of which encodes into [9F.1], and the other of which encodes into [9F.2]. Otherwise, we would be in a situation of non-fractionability (see section 8G above), which is incompatible with the hypothesis that N is a mechanism. Thus, if we want to maintain the hypothesis that N is a mechanism, we must suppose

$$f = f_1 + f_2$$

in the sense of direct sum.

But then what about Φ? As we have seen, Φ answers only the question "why f?" But the question "why f?" is now *two questions; "why f_1?"* and "why f_2?" By using exactly the same argument of fractionability, it is clear that if N is to be a mechanism, we must also be able to express as a direct sum

$$\Phi = \Phi_1 + \Phi_2$$

of two direct summands.

We thus must have the situation in which the whole graph of figure 9F3 is the direct sum of two other graphs of exactly the same form; i.e., we must be able to write

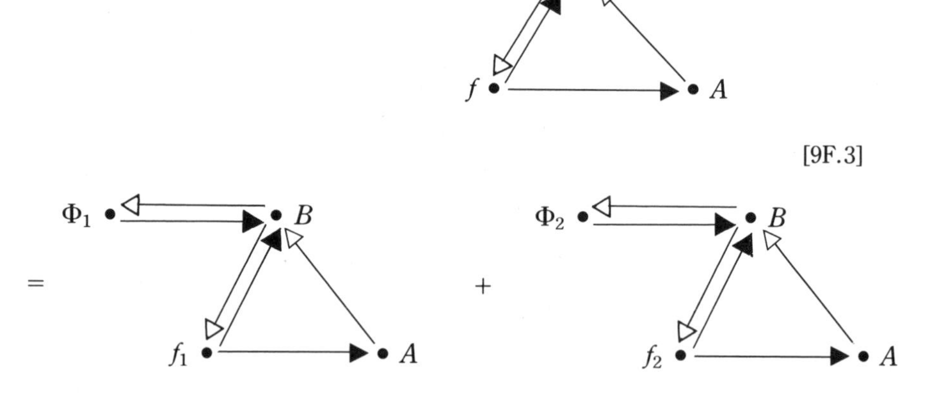

But now it is clear that we can repeat the whole argument, using the individual summands we have just extracted. Thus, the summand f_1 decomposes itself into two summands

$$f_1 = f_{11} + f_{12},$$

each of which names a disjoint set of states of *N*, etc. We thus see another incipient infinite regress forming, this time in the direction of smaller and smaller models of *N*. And obviously, this succession also violates the hypothesis that we are dealing with a mechanism.

This completes the argument.

9G. Conclusions

The preceding argument brings to a sharp focus the sprawling panorama that led up to it. It says basically that certain modes of entailment are not available in a mechanism. In particular: *there can be no closed path of efficient causation in a mechanism.* In terms of the graphs we have been using in the preceding sections, the conclusion is that *there is no cycle that contains all the solid-head arrows.*

It will be recalled that this result is itself a corollary of the profound restrictions that must be satisfied in the category of models of a mechanism. As we saw, there must be a family of minimal or smallest models, and a largest model, which is the direct sum of the smallest ones. Indeed, in this category, direct sums and direct products must coincide, and hence, we can replace an arbitrary model by a synthetic one, i.e., by a direct sum of minimal ones.

In formalistic terms, we may say that a mechanism is a system in which syntactics and semantics coincide.

Thus it is that mechanisms come to enjoy many pleasant properties, properties that precisely reflect these limitations. One can live very happily in the world of mechanisms for a long time, studying these special properties. By elevating them to Laws of Nature, we guarantee that we shall never leave this world. This is, essentially, what contemporary physics has done; it has thereby become the science of mechanisms.

If, then, there should be a material system, a natural system in the external world, that is not a mechanism, we find ourselves outside of contemporary physics. We find ourselves facing an aspect of material reality that in principle should be, but is not yet, a part of physics. Indeed, it cannot be a part of contemporary physics, precisely because the tacit epistemological presuppositions on which contemporary physics is based forbid it.

Indeed, if there should be a material system that is not a mechanism, all that contemporary physics can do about it is to tell us what properties it *cannot* have. And indeed, from its perspective, what must be absent seems devastating. For instance, such a system cannot have a state set, built up synthetically from the states of minimal models and fixed once and for all. If there is no state set, there is certainly no recursion, and hence, no dynamics in the ordinary sense of the term. There is accordingly no largest model of such a system. And the categories of causation in it cannot be segregated into discrete, fixed parts, because fractionability itself fails.

Perhaps worst of all, we lose simulability; a system that is not a mechanism must have nonsimulable models. What then becomes of prediction, of verification, of falsification, and all the other such ideas that have (mistakenly, in my view) become touchstones for science itself? Must we not then give up *science,* if we dare venture outside the realm of mechanism?

The picture we have painted looks bleak indeed, if we insist on identifying *science* with *mechanism.* But we must recall that there is no basis for such an identification. Mechanism is merely one way of expressing Natural Law, a way we happen to have become used to, to be sure, but it is itself not Natural Law, nor is it a consequence of Natural Law. Whatever happens to mechanism, Natural Law remains; hence the concept of *model* remains. It is this to which we must retreat and venture forth again in a new direction.

We already have a hint of what to do, embedded in the discussion we have given above. Natural Law tells us that if we have a natural system, we have models, formal systems whose inferential entailment structures can be brought into congruence with causal entailment in that natural system. These models still form a category, which is itself a formal object. In this category, there are generally *limiting processes* available. These limiting processes are necessarily *trivial* in the case of mechanisms but need not be so in general. Indeed we have already used such limiting processes, *and* invoked the triviality of these processes in the class of mechanisms, as an essential part of the argument we developed in the past few sections of this chapter.

These remarks provide an essential clue to how we may yet keep some of our mechanistic cake. The failure of a natural system to be a mechanism does not at all mean that it has no mechanistic models; indeed, in some sense, these form a subcategory of the category of all its models. Taking limits of these, in an appropriate judicious sense, will take us out of that subcategory but keep us within the category of models. If we are lucky, the *closure* of the subcategory of mechanical models will be precisely the whole category of all models.

In this context, the nongenericity of mechanisms becomes clear. Nongenericity, as we saw earlier, always involves some kind of degeneracy; the degeneracy here shows itself in the triviality of these limiting processes if we are already dealing with a mechanism. Degeneracy means invisibility, and this is why no amount of refinement or generalization of a mechanistic *formalism,* however radical it may appear, can do other than generate another mechanistic formalism.

Another hint comes from the existence of relational models. As we have seen, these are not tied up to any mechanistic hypotheses, and they remain meaningful even when mechanism does not apply.

These few hints suggest how it is possible to leave the world of mechanism without giving up science. Having provided them, I have thereby rendered myself superfluous; readers are now in a position to proceed for themselves and do everything that will be related henceforth on their own.

Chapter 10

Life Itself: The Preliminary Steps

THE READER who has come this far may recall that, at the very beginning, we started with a question. The question was: "What is life?" We have discussed many things between then and now, things that often seemed to ramify off in many directions unconnected with this question. But in fact, everything we have discussed in these pages is there only because it plays its role in allowing us at last to propose an answer.

10A. The Answer

The answer we propose is now this: *a material system is an organism if, and only if, it is closed to efficient causation.* That is, if f is any component of such a system, the question "why f?" has an answer within the system, which corresponds to the category of efficient cause of f. In terms of the graphs we have been using, every component must (1) initiate a red arrow, since it is a component, and (2) terminate a green arrow. We claim that everything else about organisms, everything studied in biology by biologists, and much else besides, arises from and devolves upon this property.

Accordingly, the theory of organisms, *theoretical biology,* is the study of the category of all models of such systems.

This is indeed biology from a new and different perspective. For one thing, it places the heart of biology entirely outside the scope of mechanism. This in itself is illuminating; it allows us finally to understand why the initial presumption of mechanism has made the basic question "What is life?" inaccessible and unanswerable. For another, it concretely embodies the prescient foresight of Rashevsky (see section 5B above), which initially impelled him to seek to make a relational biology in the first place.

At the same time, we enormously enlarge the scope of biology itself. Biology becomes identified with *the class of material realizations* of a certain kind of relational organization, and hence, to that extent divorced from the structural details of any particular kind of realization. It is thus not simply the study of whatever organisms happen to appear in the external world of the biologist; it could be, and in fact is, much more than that. Biology becomes in fact a *creative* endeavor; to fabricate any realization of the essential relational organization (i.e., to fabricate a material system that possesses such a model) is to create a new organism.

Seen in this light, we can see the beginnings of a *technology* that comes along with theoretical biology, a technology of *fabrication.* As we shall see in the next sections, one of the truly fascinating features of all of this is that this technology is itself described by, or realizes, the very theory that generates it.

10B. The "Machine Metaphor" Revisited

In the light of all that I have said so far, we can now see more clearly the actual status of the Cartesian identification of "organism" and "machine." In fact, our analysis will allow us to say two different things about this identification; it will allow us to specify (1) the grounds on which it is tenable and (2) the grounds on which it is not.

Let us first consider wherein the likening of "organism" and "machine" can be sustained. Primarily, this rests on the fact that *both organisms and machines admit relational descriptions.* That is to say, any organism, and any machine, can be represented in terms of an oriented graph, whose vertices are *components,* and whose various arrows represent *entailment relations* between these components. Along with these elements (namely, the components and the entailments between them) comes the crucial idea of *function,* and hence also, as we saw in sections 5I, 5K above, a perfectly rigorous notion of *final cause.*

It is these similarities that are, at least in part, immediately perceptible (at least, they were to Descartes) and provided the basis for the extrapolation he made. But it is precisely in these extrapolations that trouble arises. The whole thrust of our argument so far, in fact, is to indicate precisely that any further extrapolation from machine to organism is untenable and show why and how it is untenable.

In a nutshell, we have shown that the relational descriptions arising from machines are fundamentally different, in several ways, from those that describe organisms. For one thing, in a machine, the components themselves are direct summands of disjoint states, and the whole machine can itself be described as a direct sum of such summands. In an organism, we can make no such identification; components are not in general direct summands of anything; indeed, the concept of direct sum is not even available any more. This in turn reflects the general nonfractionability of components in an organism.

All of this in turn reflects the *impoverishment of entailment* in a machine, as compared with an organism. In a machine, entailment is in fact impoverished in two ways. First, there is the once-and-for-all segregation of causal entailment into distinct direct summands, as embodied in the fractionability I have already mentioned. This much is true in general mechanisms and is already decisive. But even more important is the characteristic absence of closed chains of efficient causation, which I have asserted in fact to be the defining characteristic of organisms.

This last is worth some further discussion, for it has always been perceived as troublesome, even by those who have lived by the machine metaphor. Specifically, suppose that f represents a component; let us ask the question "why f?" and seek an answer in terms of efficient causation: because Φ. As we have seen, if we are dealing with a machine, we cannot generally provide such an answer *within* the system; therefore, any such answer must pertain to the *environment* of the system, something outside. In effect, we are thereby ascribing a *function* to the environment, namely, to answer the question "why f?" Accordingly, we can provide an answer to the question "why Φ?" in terms of a category of *final causation:* "because Φ entails f." We thereby put not only efficient cause (of f) but also final cause (of Φ) *into the environment.*

In short, efficient causation of something *inside* the system is tied to final cause of something *outside* the system. As Voltaire once succinctly put it, "a clock argues a clockmaker," from which he then went on to conclude, by extrapolation, "*therefore* a universe argues a God." This kind of dialectic is in fact, as we have seen, inherent in the concept of a machine; once one puts efficient causation of a system component into the environment, one thereby also puts final causation outside the environment. And as I pointed out before, one of the main intentions of the machine metaphor itself was to dispense with final causation entirely. Instead, it inevitably reappears in a worse form than before.

In the context of machines, one can only obviate such problems by trying to pull efficient causation of components *inside* the system itself. As I argued in the preceding chapter, there are two different ways of trying to do this: (1) by *enlarging* the system with new states, new direct summands, and (2) by *shrinking* the system, *imposing further constraints* upon the states already present. And as we have seen, both procedures generate infinite regresses that are incompatible with the hypothesis of a mechanism.

We pause to notice that this second possibility, the imposition of more constraints, further conditions to be satisfied, takes us from something more general to something less general, from "more generic" to "less generic"; see chapter 2 above. Long ago (see section 1A above) we stated that physicists, operating always within the framework of mechanism, have long supposed that organisms were a very *special* class of material systems, in fact, too special to be of interest to them as material systems. I believe these remarks explain the heuristic basis for these presuppositions. Namely, once one starts from the presumption that a system is a mechanism, with a fixed state set and all that this entails, to say that the system is special *has to* mean the imposition of further constraints, the embedding of a sparse set of constrained states in a sea of unconstrained ones. But as I have shown, although one can get to the machines this way, one cannot get to organisms. Thus, quite the contrary is true; *mechanisms themselves are already specializations.*

Thus, we conclude that there is no way to go from machines to organisms, neither by adding states nor by subtracting (constraining) them. However, we can easily go the other way. Namely, given a relational description of an organism, i.e., an abstract block diagram of the type I have described, we can easily find subdiagrams that look like (the diagrams of) machines. Put another way, organisms generally have many different machine models. But the organism itself is not in any sense a direct sum of such models; it can only be considered as a kind of limit of them.

These same considerations also allow us to conclude that biomimesis (see 1B above) is not an effective strategy for producing organisms. As we noted, it proceeds by accretion, and hence, by direct summation. As such, it always keeps us within the class of mechanisms if we start in that class. We thus obtain only simulacra, material systems that realize some machine subdiagram of an abstract block diagram of an organism. For the same reasons, I would hazard that "artificially intelligent" systems could never be intelligent. Indeed, we can see from the above that anything like "Turing's Test" is only meaningful or applicable within the category of mechanisms; it is useless outside, where organisms are. Such tests are useless precisely because they presuppose an "exhaustion" of relational properties by machine models, something that cannot happen.

I stress again how much the relation between organisms and machines parallels the relation between natural languages and formal systems, or between formal systems and formalizations. In each of these pairs, the first member can be said to possess semantic aspects that are neither encodable into, nor decodable from, the latter. In fact, it is not too far wrong to say that an organism, in the sense we are employing the term, is itself like a little natural language, possessing semantic modes of entailment not present in any formal piece of it that we pull out and study syntactically.

Let us then sum up the status of the "machine metaphor." It succeeds in likening organisms and machines, to the extent that both classes of systems admit relational descriptions. But beyond that, it is fundamentally incorrect; it inherently inverts the notions of what is general and what is special.

On balance, the Cartesian metaphor of organism as machine has proved to be a good idea. Ideas do not have to be correct in order to be good; it is only necessary that, if they do fail, they do so in an interesting way.

10C. Relational Models of Organisms

We are rapidly reaching the place where we will stop for the present, a rather high plateau where we may all rest before pushing further. It is the function of the present section to get us securely onto that plateau.

We have already left states, and recursive state transitions, behind us. In their place, we have abstract block diagrams and chases through these diagrams. As we have seen, these diagrams embody causal relationships; they allow us, from their structure, to read off the answers to questions of the form "why?" about the relational constituents of the system. Sometimes there are no answers to be found within the system, but that can be determined from the abstract block diagram as well. In such a case, the missing functions, (i.e., the ones not available *within* the system to answer "why?" about other elements of the system) must be ascribed to the environment. As we have seen, such abstract block diagrams can be conventionally abbreviated as a diagram of sets (or, in the neutral language of category theory, of objects) and mappings (morphisms) between them.

When we do have states available (i.e., if we are dealing with mechanisms and machines), then, as we have seen, each element in such an abstract block diagram is a name for a certain set of states, which constitutes a direct summand. In such a case, the causal structure encoded into the abstract block diagram arises from the recursive state transition sequence in the system as a whole; as we recall, this is all the causality that there is in mechanisms. In that case, the abstract block diagram represents a system of *constraints*. But as I showed in the preceding section, the abstract block diagrams that can arise from such a situation are extremely limited, especially in terms of efficient causation. Most of the "why?"

questions we can ask about such a system are unanswerable within the system, and therefore, must be referred to its environment. Put another way: most elements of an abstract block diagram arising from a mechanism are *unentailed.*

My claim is that organisms lie at the other extreme as far as entailment is concerned. Their abstract block diagrams manifest *maximal entailment;* in particular, if f denotes a component of such a system, the question "why f?" has an answer, in terms of efficient causation, *within the system.* Unfortunately, this much entailment is simply not compatible with the idea of mechanism; therefore it can no longer be interpreted or understood in purely syntactic terms. Organisms then must manifest an inherently semantic character when viewed from the standpoint of mechanisms. As with the parallel situation of formalization in mathematics, this fact only gives trouble if we insist on believing that syntax exhausts the world.

Let us now proceed to illustrate this last circle of ideas, by looking at a specific situation and in effect successively enriching it in entailment, pulling all the entailments we need inside. We have already seen much of this before (see section 5J above), all but the last step.

So let us start with the simplest situation, a single component f, which gives rise to a relational diagram of the form

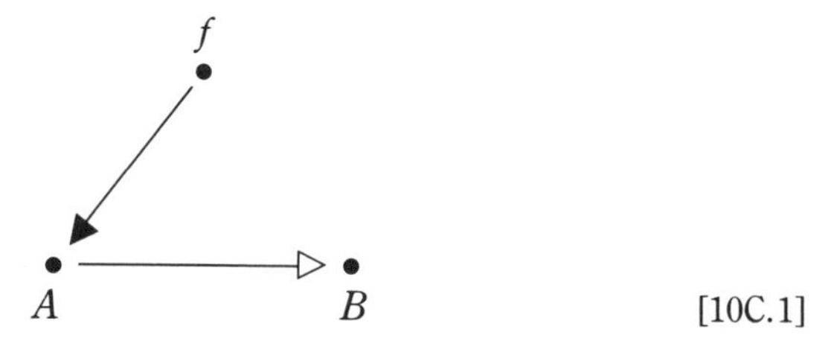

[10C.1]

or, as we customarily abbreviate it,

$$A \xrightarrow{f} B \qquad [10C.2]$$

As we saw, this diagram may (although *it need not*) itself be interpreted in completely mechanistic terms. To adopt a provocative terminology, I refer to such a diagram as *metabolic;* it involves merely a processing, which produces outputs in B from inputs in A.

In such a diagram, the processor f is itself unentailed. As we saw, one way to entail it is to enlarge the diagram, to throw in a new processor Φ, a new mapping, which intuitively makes f from an appropriate set of inputs X. For maximum economy, we may take $X = B$; why not? Then, as we saw, we get a bigger graph of the form

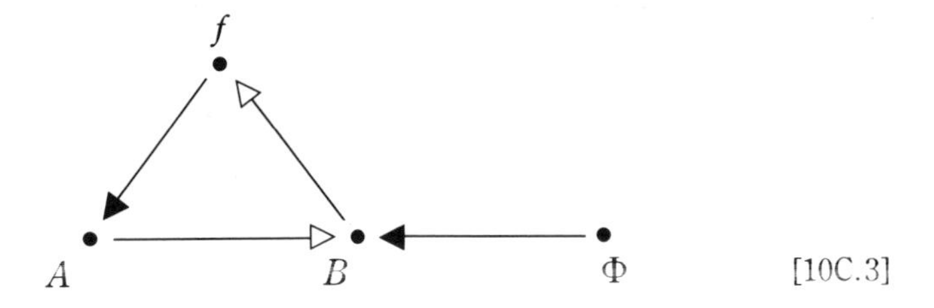

[10C.3]

which we can correspondingly abbreviate as

$$A \xrightarrow{f} B \xrightarrow{\Phi} H(A, B). \qquad [10C.4]$$

In this larger relational description, the mapping f, what we called the "metabolic processor," is now entailed by something. Thus, the graph depicted in [10C.3] or [10C.4] is the original metabolism plus something, some new function. Let us call that function *repair.* Thus it is natural to call any system, which has a relational diagram of the form [10C.3] or [10C.4], and hence, embodies the two distinct functions of metabolism and repair, an *(M, R)*-system.

But now Φ is unentailed, as we have seen; we do not yet have enough entailment in the system to answer the question "why Φ?" Intuitively, if we did have it, then we would thereby embody *another function,* in addition to the two (namely, metabolism and repair), which the diagram already manifests. Let us give a name to this new function, which would entail repair; let us call it *replication.*

A simple-minded iteration of the preceding argument would lead us to throw a new vertex β, embodying this new replication function, into the graph and draw the appropriate arrows. Again assuming that the domain of B is to be identified with the range of Φ, we get

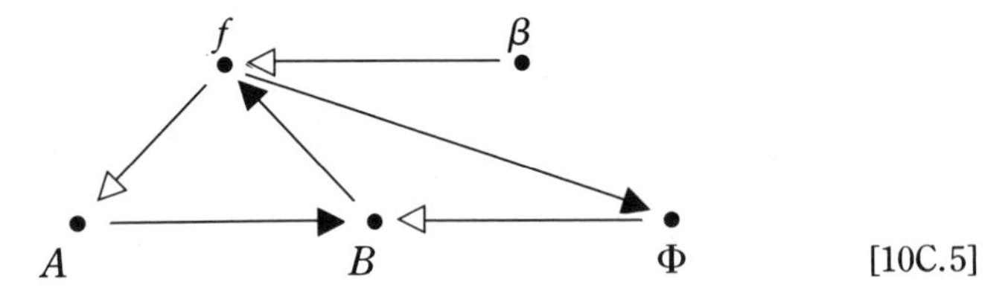

[10C.5]

or, in abbreviated form,

$$A \xrightarrow{f} B \xrightarrow{\Phi} H(A, B) \xrightarrow{\beta} H(B, H(A, B)).$$

But once again, this seems to gain us nothing, since the new vertex β, representing the new replication function, is unentailed.

The point of departure for everything I have said in the present volume in fact arose from my discovery, a very long time ago (in 1957, to be exact) that β was not a vertex. In effect, the function of replication could already be represented simply by throwing more arrows, more entailment, into the diagram we already had. In other words, the *function* of replication could be, under certain formal circumstances, already entailed by the two prior functions of metabolism and repair.

I have since repeated this formal argument many times in previous work and need not repeat it here. In brief, it involves two steps: first, to regard an element b in B as itself a processor; as a mapping $\hat{b}$ defined by

$$\hat{b}(f) = F(b).$$

This is not yet the mapping we really want; what we want is

$$\hat{b}^{-1} \equiv \beta$$

if it exists. Readers can readily establish for themselves, or go back to the earlier published versions of the argument, that the condition for invertibility of $\hat{b}$ is really a condition on $H(A, B)$. If that condition (which is restrictive, but not unduly so) is in fact satisfied, then our diagram becomes

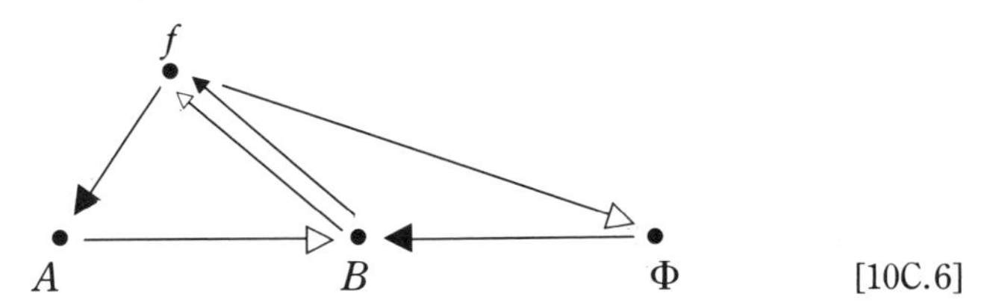

[10C.6]

In this graph, every function is indeed entailed by another function in the graph itself. As far as entailment is concerned, the environment is out of the picture completely, except for the initial input A.

It goes without saying that this graph possesses a curious and interesting structure, as compared with anything we have seen before. In particular, we cannot abbreviate it in terms of sets and mappings without seriously misrepresenting its properties. From this perspective, we can see why the peculiar-looking notation we have adopted was in fact necessary.

The reader may note that there is a certain parallel between the invocation of replication above, and Newton's Second Law. Both serve to stop an incipient infinite regress, in each case by making what happens at an early step in the regress depend on preceding steps. For Newton, this involved identifying acceleration as a function of phase; for us, it is expressing the function of replication in terms of the functions of metabolism and repair.

Clearly, the diagram [10C.6] satisfies the condition we have laid down at the outset of this chapter, namely, that everything in it is entailed in the sense of efficient cause entirely within the diagram. Any material system possessing such a graph as a relational model (i.e., which *realizes* that graph) is accordingly an organism. From our present perspective, we can see that [10C.6] is not the only graph that satisfies our conditions regarding entailments; there are many others. A material realization of any of them would likewise, to that extent, constitute an organism.

I am asserting, then, that biology as a science is concerned with those material systems that realize an abstract block diagram like [10C.6]. Some of the properties of any such system arise entirely from the fact that it realizes the diagram; others arise from *how* it realizes the diagram; others will depend on both. All of these are, of course, embodied in the structure of the category of all of its models. Thus, from our present vantage point, biology comprises two dual aspects: (1) the class of all material systems that realize a certain kind of relational structure, and (2) given any such, the structure of the category of *all* of its models. The two together embody what Rashevsky called "the unity of the organic world as a whole." It is a large and daunting thing.

10D. A Word About Fabrication

Given a component *f*, the entailments it generates (the flows it specifies from inputs to outputs) constitute *physiology*. But when, as we have done, we look upon *f* itself as *effect* by asking the question "why *f*?" we are in a quite different, and generally unrelated causal world. Ironically, the only answer to the question "why *f*?" that pertains directly to the physiology of *f* belongs to the category of final causation; "*f* because *f* entails. . . ."

This world is the world of fabrication, and we have already been inside it for some time; it is perhaps the essence of organism. In the *(M, R)*-systems we have just described, the component Φ, which embodies the function we call repair, is a fabricator of *f*. The physiology of Φ is precisely the fabrication of *f*, and the *(M, R)*-system itself describes a situation in which the fabrication process itself has been pulled inside the very system in which what is fabricated is operating.

Every *(M, R)*-system is thus, in itself, a little theory of fabrication. To realize an *(M, R)*-system is to produce a kind of factory, within which another kind of factory is itself entailed. Conversely, whenever we pull a fabricator inside a system by putting it together with what it fabricates, the result is essentially an *(M, R)*-system.

If we have such a system (call it *S*), we can obviously ask "why *S*?" This is the *origin-of-S* problem. When we have solved it, we have thereby in effect produced a new *(M, R)*-system, with *S* itself as metabolic part.

If *S* is a *replicating (M, R)*-system, one obvious answer to the question "why *S*?" is the circular one, "because *S*." In this situation, the physiology of *S* already entails enough fabrication to answer the question. In other words, one *(M, R)*-system that answers the question "why S?" happens in this case to be *S* itself.

If this were the *only* way to answer the question "why *S*?" then the origin-of-*S* problem has, in material terms, no real solution; if *S* exists in this situation, it does so because the universe is what it is; *S* has no other efficient causation than itself. *S* thus acquires a status akin to the elementary particles of the physicist, which likewise admit no category of efficient causation.

Otherwise, we can embed *S* into a strictly larger system, which is an *(M, R)*-system, in such a way that the operation of the larger system serves precisely to fabricate *S*. Hence to realize this larger *(M, R)*-system is to produce a fabricator of *S*, and hence, to solve the origin-of-*S* problem.

Thus, we can see that a *theory* of fabrication is already intimately tied to the concept of an *(M, R)*-system. It becomes especially interesting when the system to be fabricated is itself an *(M, R)*-system (and in particular, a replicating *(M, R)*-system). From our present plateau, we can see this next peak to be scaled, looming above us. But we are in a good position now from which to launch an ascent; all the tools are now in our hands. Indeed, readers do not have to wait upon me; from where they stand now, they can set out on this journey for themselves, any time they care to.

Chapter 11

Relational Biology and Biology

WE BEGAN this discussion with the question "What is life?" We ended with the answer: Life is the manifestation of a certain kind of (relational) model. A particular material system is *living* if it realizes this model.

11A. What Is Biology?

I have always regarded biology as the study of the living; i.e., of the specific instantiations or manifestations of life. The basic operational question for biology has always been: how can we learn about this life through a study of the living? How can we extract from these samples or specimens of life an *essence* they all share?

I thus anchor biology entirely in what August Weismann called the *soma.* The soma is what is alive; the heart of what constitutes organism. What I have argued above, at great length, is that this soma, in its operations, in the way it realizes the underlying relational model, cannot be a machine. That is, it cannot be expressed in terms of mere syntax; there has to be too much entailment in it for that. In a word, it must be *complex.* We cannot answer the question, "What is life?" or the question "Why is an organism alive?" with the answer "Because it is a machine." If we try to simplify in this way, we must inevitably lose most of the entailment we need; we literally kill life.

Nor can we answer the question "Why is an organism alive?" with the answer "Because its ancestors were alive." Pedigrees, lineages, genealogies, and the like, are quite irrelevant to the basic question. Yet they are the very stuff of evolution. Ever more insistently over the past century, and never more so than today, we hear the argument that biology *is* evolution; that living things instantiate evolutionary processes rather than life; and ironically, that these processes are devoid of entailment, immune to natural law, and hence outside of science completely.

To me, it is easy to conceive of life, and hence biology, without evolution. But not of evolution without life. Thus, evolution is a corollary of the living, the consequence of specialized somatic activities, and not the other way around. Indeed, it may very well be more a property of particular realizations of life, rather than of life itself. Thus it is that the word "evolution" has hardly been mentioned in the preceding pages.

There are thus not one but two radical biological correlates of what we have done above. The first is an assertion that biology is wrapped up with soma and how it operates; thus we cannot invoke evolution as an explanatory or causal principle for these purposes. To cite once again the words of Laplace: "Je n'ai pas besoin de cet hypothèse." On the other hand, if this much is true, it is also true that evolution itself is entailed; it is far from the whimsical history of "frozen accident" of current belief.

The second radical thing is a denial that soma is a machine in any sense, i.e., something to be characterized through reductionistic modes of material analysis. What we capture that way are properties of how life was realized; not of life itself. As we have seen above, this has radical consequences not only for biology, but for the physics on which it is presumed to rest.

In what follows, I will enlarge on some of these assertions, by looking at these facets of contemporary biological beliefs from the perspective we have developed above.

11B. The Paradoxes of Evolution

Evolution has, as we have seen, come to do for biology today what vitalism did for it previously. Vitalism, in its most general form, simply asserted that whatever made organisms alive was forever out of the reach of what governed inanimate nature. Some additional principle, some vital force, had to be invoked; it was precisely this additional principle, missing from the rest of nature, which made organism *in principle* inexplicable via inanimate nature alone.

In the past, the only perceptible alternative to this vitalism was mechanism; the converse claim that there is *nothing* in biology which cannot be understood in precisely the same terms in which anything else in nature is understood. As we saw at the outset, the Cartesian Machine Metaphor offered precisely this; a way of anchoring biology in science by de-mystifying it, and subsuming it entirely into physics. By physics, of course, I mean the *contemporary* physics of the time; whether in 1650, or in 1950, or today.

Both of these positions have had over the years an immense allure to the biologist. Mechanism offers a secure place for biologists in science, as scientists among other scientists; respectability through the adoption of a common, shared tradition and shared principles. They need no longer be dismissed contemptuously as "birdwatchers" or as mystics. It is very important to biologists to share in this security; to be "inside."

On the other hand, biologists have always liked to believe that what they study *is* somehow different, and hence that they are not just like everybody else. The cost of espousing mechanism was precisely that they give this up. As long as the only alternative seemed to be vitalism, they predominately, though grudgingly, chose the former.

Biologists today have come to see in Darwinian evolution a way of distinguishing themselves again, of making themselves separate, without the vitalistic traps. Basically, the argument is now that it is evolution which is unpredictable, non-mechanical, immune to the entailments, the causality, the determinism which mechanism made them espouse. By the single, simple act of *redefining* biology, to assert that it is about evolution rather than about organism, we can in effect have our mechanistic cake, and eat our vitalistic one too. Biologists continue to espouse a most narrow form of mechanism as far as what goes on *within* organisms is concerned. But if biology is about evolution, these mechanistic shackles can be devalued; conceptually assigned a subordinate role. One can (at least apparently) embrace evolution without having to deny mechanism; but we can thereby devalue it.

To avoid making evolution subject to mechanism is therefore essential. But it is also essential to avoid asserting anything vitalistic. The only way to do this is to deny any vestige of entailment in evolutionary processes at all. By doing so, we turn evolution, and hence biology, into a collection of pure historical chronicles, like tables of random numbers, or stock exchange quotations.

This absolute denial of entailment in evolutionary processes is thus a central, perhaps *the* central pillar, of the current biological *weltanschauung*. If we *did* admit entailment into the evolutionary realm, then only two alternatives seem visible: (1) these entailments are themselves mechanistic, in which case biology disappears back into mechanism again, and loses forever its distinguished character, or (2) these entailments are not mechanistic, which seems to mean they must be Vitalistic again. Both of these, for different reasons, are quite unacceptable. Hence we are *driven* to expunge entailment from evolution entirely, not on any intrinsic scientific grounds, but because of the psychological requirements of biologists.

This picture struck me early as a kind of mythology, with evolution as protagonist, in its exact dictionary meaning of "serving to explain or sanctify some concept, usage, institution or natural phenomenon."

It was, for instance, entirely on such grounds that the ideas of Walter Elsasser, which we briefly discussed earlier (section 1A), were not only dismissed, but violently attacked, by those biologists who bothered to read what he had written. All Elsasser did was to exploit the assumption that organisms are "rare" among material systems, and hence disappear from "general laws" obtained by averaging. He was thus led to suggest that, in the sparse realm he envisaged for biology, there would be "biotonic laws" governing what went on there; not derivable from the average, "general" laws, although compatible with them. It was this last suggestion, that "laws" were operative at all in this biotonic realm, which exposed him to violent attack from the biological side. Mere paraphrase cannot convey the character of these. Here, for instance, are the words of Jacques Monod:

> Summarized in a few words, here is Elsasser's position.
>
> The strange properties (of organisms) are doubtless not at odds with physics; but the physical forces and chemical interactions brought to light by the study of nonliving systems *do not fully account for them.* Hence it must be realized that over and above physical principles *and adding themselves thereto,* others are operative in living matter, but not in non-living systems where, consequently, these electively vital principles could not be discovered. It is these principles—or, to borrow from Elsasser's terminology, these "biotonic laws"—that must be elucidated. . . . The least one can say is that the arguments of these physicists is oddly lacking in strictness and solidity. (*Chance and Necessity,* pp. 27–28; emphases in original).

With this language, then, Monod consigned Elsasser to the category of "scientific Vitalism," one of the lower rungs in his scientific Hell. And yet, all Elsasser did to deserve this was to draw an inconvenient conclusion from Monod's own assertion, embodied in the first few sentences of the preface to *Chance and Necessity,* that "Biology . . . (is) marginal because —the living world constituting but a tiny and very 'special' part of the universe—it does not seem likely that the study of living beings will ever uncover general laws applicable outside the biosphere."

Monod's language, and that of countless other similar assertions which could be adduced, is clearly not the language of collegial scientific discourse. It is rather the response of someone who feels his myths are under attack. That is, it expresses a religious rather than a scientific attitude.

This, then, is the paradox of evolution—that it has come to play an essential mythological role in the world-picture of contemporary biologists. It was initially regarded as a way to bring the panorama of biological species into the realm of science. Its present role is, rather, to excuse itself from science through its absolute denial of evolutionary entailments, and thereby to consign itself (and hence biology, as presently constituted) to the realm of historical chronicle.

However, as we have abundantly argued in the preceding pages, it is not a question of mechanism or vitalism. It is a question of simplicity or complexity.

11C. Mendel, Heredity, and Genetics

Evolution is concerned with "explaining" the characteristics of present generations retrospectively, in terms of the characteristics of temporally remote ones. Heredity is concerned with what passes between one generation and the next. Thus, if we view evolutionary processes as a kind of time integral of hereditary ones, in which a "generation" provides the individual time-step by which evolutionary chronicles are indexed, it is important to have a tangible bridge between today's generation and yesterday's, the immediately preceding one. This, roughly, is where *genetics* enters the picture.

Let us begin with a few historical comments.

Gregor Mendel was perhaps the first to concern himself with how the forms and behaviors of parents are related to those of their offspring. That is, he was comparing *phenotypes,* embodied in particular phenotypic *characters,* between one generation and the next. It seems to me, although as far as I know Mendel did not say so, that he was much influenced by the chemistry of his time. In chemistry, a small number of different kinds of atoms, the chemical elements, could enter into an unlimited number of *compounds.* These compounds could be of utterly diverse *character,* or "phenotype"; they could be solid or fluid, crystalline or amorphous, red or yellow. In any case, the underlying atoms could be retrieved unchanged from any such combination into which they had entered, and recombined to yield new compounds, with new "phenotypes." Chemical reaction, in its turn, could be regarded as a way of liberating the constituent atoms and recombining them, and thus turning given "phenotypes" into new ones. This clearly constitutes a powerful metaphor for what goes on between parents and the production of offspring. Note carefully that this viewpoint does not, in itself, constitute a *reduction* of heredity to chemistry; the relation between them is a metaphoric one, one of alternate realization. But it is powerful nevertheless.

Let us explore this metaphor a little further. In particular, let us ask whether the atomization of chemistry in any way "atomizes" the phenotypes manifested by individual chemical compounds in bulk. For instance: are there "atoms" for crystallinity, or solidity, or color, or any other such morphological characters? Even in chemistry, it is not nearly as simple as that; the atomization of a molecule does not, in that sense, at all entail a corresponding "atomization" of its phenotypic characters. Nor does it entail any kind of 1 to 1 mapping between these characters and those of the atoms, or even the molecules, which underlie them. Put in another language, which we developed earlier, the phenotype cannot be *factionated* into "atoms" as a constituent molecule can; nor can the actual atoms be fractionated into separate phenotypic characters for which they are responsible. We will come back to this, or at least to its hereditary analogues, in a moment.

Returning to Mendel: he published his evidence for atom-like, particulate *factors* (as he called them) in heredity in 1866. As is well known, his papers sank without a trace. The fact is that no one then cared much about heredity *per se;* it was not until the connection between heredity and evolution, about which everybody cared, was slowly perceived, that Mendel's ideas were retrieved from oblivion. It was only then that the study of heredity, soon to be rechristened *genetics,* became a crucial part of the biological main stream. By that time, Mendel himself was long dead.

Implicit in Mendel's ideas was a duality between the phenotypic *characters,* manifested in the forms and behaviors of individual organisms within a generation, and the particulate *factors* which pass between genera-from parent to offspring. This is the genotype-phenotype dualism, which was first stated by August Weismann in his influential book *Das Keimplasm,* published in 1874, and to which I have already alluded. He proposed, in apparent ignorance of Mendel's work, a duality between what he called *soma* and what he called *germplasm.* Roughly speaking, the somatic part pertains to what is mortal in biology, with what we today identify with phenotype. Germplasm, on the other hand, pertains to what is immortal, flowing from generation to generation from the beginning of life on the planet. Thus the soma must be recreated anew in each generation, in such a way as to facilitate the flow of the precious germplasm. This was Weismann's conception of evolution itself.

We have already alluded to this dualism in the preceding section, where we identified organism with soma, and with life itself. Weismann, on the other hand, was the first to propose identifying life with germplasm, and with its flow. He thus, if indirectly, provided the crucial ingredient for the outlook embodied in contemporary biology.

But, as we have seen, Mendel himself was concerned primarily with this soma, with phenotype. He characterized his particulate factors in terms of their *functional* roles in this regard; i.e., in terms of how they manifested themselves in somatic forms and their generation. That was how his factors were recognized, and in turn, that is what the factors were *for*.

A basic transformation of the situation occurred roughly between the years 1900–1910. During these years, the identification came to be made between the hypothetical Mendelian "linkage groups" and the tangible, material, cytological structures called chromosomes. Accordingly, the Mendelian factors themselves, the genes, must be materially incarnated somehow in the *structure* of these bodies. So instead of resting content with characterizing these genes in *functional* terms, through their modulation of form and morphogenesis, we could rather imagine characterizing them *intrinsically,* in independent *structural* terms. That would, if accomplished, leave only the problem of accounting for the genes' functional, morphogenetic activities in terms of this basic, intrinsic structure.

All this has a very contemporary ring; indeed, we are almost at the molecular biology of today. And of course, it is all fully in accord with the basic machine metaphor; chromosomes could be fractionated from cells, and genes from chromosomes, purely as material structures, as independent material systems, without loss of information.

At the same time, the functional relation between genotype and phenotype was becoming progressively more complicated. Mendel thought that each of his functional factors could be associated with a fixed, demarcated phenotypic character; the one-gene-one-character hypothesis. As we have already suggested, it is not that simple; the dualism between genotype and phenotype does not extend to any kind of 1 to 1 mapping between them. Geneticists early discovered phenomena of *pleiotropy,* in which the "same gene" is involved in many phenotypic characters; and inversely, of *polygeny,* in which the same character is associated with many different genes. On top of this, there was growing evidence, soon to become an absolute necessity, for "cross-talk" among the genes themselves, directly or indirectly. All this was evidence of *nonfractionability,* but being uncongenial to a machine metaphor, it was regarded as a mere technical inconvenience, to be subsumed into a bigger machine.

All this, it will be noted, pertains primarily to Weismannian soma. What about evolution, which was regarded as the primary issue? Here, the genotype-phenotype dualism manifests itself in another form. Namely, the operation of Darwinian selection pertains to phenotypes; to soma. On the other hand, evolution itself pertains to the flow of genotypes; to germ-plasm. Thus, we must regard selection on phenotypes as a modulator of this flow of genes from one generation to the next. This is where the notion of *fitness* comes in.

As presently viewed, fitness involves a decision made by natural selection about a particular soma, a particular phenotype. It is a decision that can be *imputed* to the associated genotype. A low fitness rating translates operationally into a disadvantage in populating the next generation; in leaving offspring. Thus, a low somatic fitness serves as a filter, which prevents the associated genotype from reaching the next generation. Thus the gene flow is modulated, and the "gene pool" will manifest itself in somatically fitter individuals than we started with.

On the other hand, there are no "genes" for fitness either. Indeed, it cannot be regarded as a somatic feature in the usual sense at all. Fitness can only be defined operationally and retrospectively as far as individual organisms are concerned. In particular, it cannot be fractionated from anything. It is in fact a very mysterious concept; evolution in the Darwinian sense would be unthinkable without it, but it has always given evolutionists the greatest difficulty.

11D. Biochemistry, Genetics, and Molecular Biology

As we have seen, the introduction of the concept of *genotype* amounted to posing a new causal category for talking about somatic effects. That is, we could now answer a question of the form "why this somatic character?" with an answer "because these genotypic factors."

Biology today is furthermore completely committed to the idea that *phenotype is a chemical concept.* In other words, biological forms, and the morphogenetic processes which generate them, are ultimately chemical, or biochemical in nature. This, it will be seen, is a far cry from asserting an *analogy* of the kind sketched in section 11C above. This is a *reduction.*

On the face of it, it is an astonishing claim. It is not adduced on the grounds of great success in faithfully translating anatomical, embryological, or physiological processes into a syntactic chemical language. Quite the contrary, in fact, is true. It is adduced, rather, primarily on the grounds that otherwise, we simply could not answer "why?" questions about these processes with a "because . . ." framed exclusively in terms of *intrinsic,* fractionable (i.e., chemical) structure. That is, unless we *identify* phenotype with biochemistry, we can no longer claim that the functional genetic factors originally posited by Mendel can be identified with fractionable pieces of chromosomal structure. Indeed; the viability of the entire machine metaphor in somatic biology currently rests precisely here; on the identification of phenotype with biochemistry. If it fails, then mechanism fails; but as we have stressed many times before, the alternative is not vitalism, it is complexity.

Historically, the actual relevance of biochemical analyses to biology itself was open to serious question for a long time. Well into the present century, most biologists regarded the biochemists' starting point, a homogenate or slush, as artifactual to begin with, a "mere" abstraction, with no more relation to life than leather has to skin. But gradually, under the hands of its most skillful practitioners, beginning perhaps with Pasteur, certain kinds of general patterns began to emerge. It became meaningful, for instance, to talk about "the" citric acid cycle, in sources as diverse as yeast and pigeon breast muscle. An even more general pattern was found in the role of enzymes, proteinaceous catalysts, which precisely modulated individual reactions in cells. Indeed, insofar as a cell can be specified in terms of the specific reactions going on within it, cells can be characterized in terms of the kinds and amounts of enzymes which are present. From this, it was only a short step to identifying enzyme synthesis, protein synthesis, with the functional role of genome itself.

Perhaps the actual origin of what was then called "biochemical genetics" lies in the experiments of Beadle and Tatum on the fungus *Neurospora* around 1940. On the basis of their observations on nutritional deficiency in mutants, they posited a "one-gene–one-enzyme" hypothesis; a more defensible version of the earlier one-gene–one-character hypothesis we alluded to earlier. This marked, I believe, the true starting point for both the identification of enzymes with phenotype, and the synthesis of enzymes, and of proteins in general, as the precise functional role of genome in general.

People had long been looking for a fractionable chemical constituent of chromosomes, to which this function could be assigned. Attention gradually became directed toward a phylum of copolymers collectively called DNA's. These were copolymers formed by the condensation of certain nucleotides. They initially did not look very interesting, and were consigned a purely structural role, much as collagen proteins play elsewhere in the organism.

The situation was transformed dramatically with the publication of the Watson-Crick model for DNA in 1953. Attention quickly focused on the *sequence* in which the constituent monomers were arranged, and the possibility of a purely syntactic cryptographic relation between a sequence of nucleotides in DNA and a sequence of amino acids, the primary structure, of an associated polypeptide. Moreover, an equally syntactic relation could be perceived for replication, since the Watson-Crick model was a double-stranded structure. Further detail here is quite superfluous. Suffice it to say that the functional gene of Mendel has been identified with DNA structure, and more specifically, with *sequence.* This is the content of the *sequence hypothesis.*

Sequence thus becomes an answer to a question, "why does this protein have a particular primary structure?" and thus constitutes *information*—genetic information. Moreover, it is difficult not to further regard this "genetic information" entirely as software, as *program,* to protein hardware.

On the other hand, we must protect evolutionary processes from entailment; i.e., free them from program. This is accomplished by the central dogma of Crick, dating from 1957. This says basically that the outputs (polypeptides) cannot change the machinery which generated them; i.e., that protein cannot change DNA sequence *in the same generation.* This serves to liberate DNA sequence (which we have identified with "genetic information" itself) from anything in the system which might entail its modification. Stated otherwise, the only such modifications allowed arise from *errors;* unpredictable and extraneous departures from program; unentailed and unentailable. These in turn create the variability necessary to drive natural selection and hence evolution. Therefore, evolution remains itself unentailed and unentailable.

This picture is recommended by its apparent simplicity. But it becomes ever more complicated to maintain this simplicity in detail. Like all syntactical schemes, it is elastic enough to accommodate any individual new circumstance. But in this case, as we have seen, it depends on an absolute identification of phenotype and conventional biochemistry, because it mandates precisely that *all* the genes can do is already biochemical. Or, to put it contrapositively: if any aspect of phenotype cannot be expressed biochemically, then these genes cannot account for it.

11E. Chemistry and Sequence

In this section and the next, we will consider the identification of somatic phenotype with chemistry, as mandated by the sequence hypothesis. Our conclusion will be that it is false, unless chemistry itself is redefined. Ironically, it is the concept of "sequence" itself which will indicate how this redefinition, or generalization, has to take place.

Let us begin with the structural formulas of chemistry, which schematize the arrangements or dispositions of chemically bonded atoms in space, the structure of molecules. Mathematically, these take the form of connected oriented graphs, with atoms as vertices, and chemical bonds as edges. In a certain sense, these graphs, or structural formulas, are generalizations of *words,* the elements of free monoids $A^{\#}$ on finite alphabets A. That is to say, graphs are syntactic objects, but with more syntax than words have. In a sense, the analogs of concatenation in graphs extend into two or more dimensions, whereas words admit only one. Consequently, to express a graph (e.g., a structural formula) as a linear word poses some very nasty problems.

Historically, structural formulas became important because molar expressions like $C_{12}H_{22}O_{11}$ were simply too ambiguous; many chemically different molecules *(isomers)* can answer to the same molar description. Graphical or structural formulas are a way of lifting this ambiguity.

In polymeric molecules, built out of a fixed family of concatenated monomers, such as DNA molecules, or polypeptide chains, one manifestation of isomerism is precisely in the order, or sequence, in which the constituent monomers have been concatenated. From this viewpoint, sequence, expressed again as a linear word, is simply a shorthand for a much more complicated graphical structure, which itself is connected, and extended, in more than one spatial dimension.

In order for this shorthand itself to be meaningful, several things are necessary. First, *there must be no entailment between the symbols that enter into it.* This means that the symbol next to a given one must not be entailed by the given one; it must be independent of it. Thus, in a polypeptide chain, the "next" amino acid in a molecule must not be entailed by the "present" one, or by the entire chain from the beginning to the "present." Thus, any concatenation of these symbols must be an allowable word, and therefore must represent a real chemical structure. Put otherwise: the entailments which actually order such sequences must be entirely *outside.* As a *chronicle* (i.e., as monomers indexed by coordinates) we must have a complete absence of recursivity.

So there can be no *intersymbol influences.* Most importantly: we cannot *hybridize* these symbols to create new ones. Schematically, if the symbols I and O are elements of our alphabet, we cannot allow things like

$$I + O = \mathrm{G} + \reflectbox{G}$$

This would be tantamount to saying that there is internal structure in our symbols, which can manifest itself in the creation of *new symbols and new syntax.* Not only would this violate the shorthand, it would defeat the purpose of our syntax in the first place.

On the other hand, the formalism of graphs and words used to express chemical structure, however much it may resemble formalization, differs from it in one essential respect. In syntactic formalisms, the alphabet symbols are presumed *meaningless* (i.e., structureless) in themselves; they possess no external referents, and only their concatenations can be assigned meaning. In chemistry, on the other hand, the alphabetic symbols represent *atoms*. The central presupposition of reductionistic analysis is that the concatenations of these, the words formed from them, derive their meanings or properties *only* from these atoms or elements. It is the clash of these last two assertions which, in the last analysis, makes a purely syntactic approach untenable.

By virtue of illustration of these ideas, let us look at the relation of individual amino acids to their plymers, the polypeptide chains.

An individual amino acid has a generic structural formula of the form

```
H   R       O
|   |     //
N — C — C
|   |     \
H   H       OH
```

Here, R is an *abbreviation,* a shorthand, for one or another chemical groups, which itself has an extended structural, graphical formula. Whatever such group replaces R determines which specific kind of amino acid we have. Thus, if R = H, hydrogen, we are talking about *glycine;* if

```
              |
          H — C — H
              |
              C = C — H
              |     \
R =           |       N — H
              |     /
              C = C
            /       \
      H — C           C — H
            \\      //
              C — C
              |   |
              H   H
```

we are talking about tryptophane. As is well known, there are about twenty different such groups, any of which can replace the generic R. The funda-

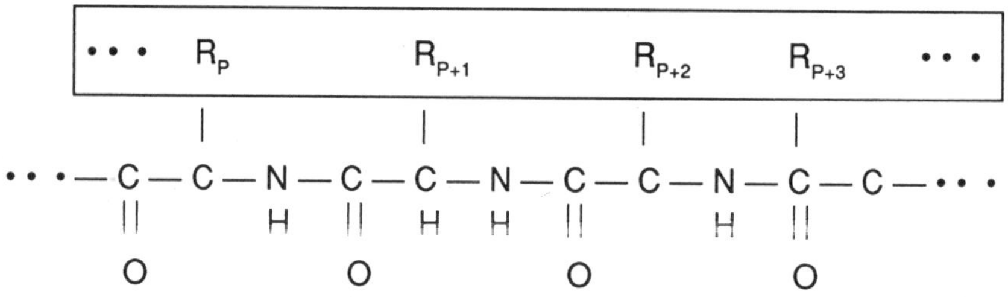

```
        ┌──────────────────────────────────────────────────────────┐
        │ • • •   Rp          Rp+1          Rp+2      Rp+3   • • • │
        └──────────────────────────────────────────────────────────┘
                 |            |              |         |
• • •—C—C—N—C—C—N—C—C—N—C—C—• • •
      ||    H  ||  H  H  ||       H  ||
      O        O         O           O
```

FIGURE 11E.1

mental concatenation step, which makes two monomers into a dimer, is a chemical condensation between the amino (NH_2) group of one of them and the carboxyl (COOH) of the other to split out water. In these terms, the reaction is

$$-NH_2 + HOOC- \rightarrow -\overset{H}{N}-\underset{\underset{O}{\|}}{C}- + H_2O.$$

The first product is essentially the *peptide bond.* Note that its formation is quite independent of the *R*'s, and that it leaves us with something which, like the amino acids themselves, still has a free amino group at one end, and a free carboxyl group on the other. Thus this concatenation operation can be iterated ad lib.

So far, all is in order. A polypeptide chain is thus something of the form figure 11E.1. In this figure, I draw your attention to the *sequence*

$$\ldots R_p R_{p+1} R_{p+2} R_{p+3} \ldots$$

which is displayed in the box. This constitutes the *primary structure* of the polypeptide, and as we stated, constitutes a syntactic shorthand for the molecule as a whole. Note that the same kind of thing can be done for any other kind of copolymer, such as DNA or RNA.

Now in inherent chemical, structural terms, what is this sequence? As it stands, it is a subgraph of the structural formula of the whole polymer molecule. But there is something *peculiar* about it, viewed in those terms. Namely, it is a *disconnected* subgraph. It is not held together by internal chemical bonds *of its own.* It thus cannot represent, on its own, a *molecule* of anything. On the other hand, it looks like a molecule, and as we shall see, it can *act* like a molecule. It is even, in some sense, a *part* of a conventional molecule. What we shall focus on, especially in the next section, is: what *kind of part* is it?

I am beginning here to draw a distinction between a polypeptide chain as a conventional chemical molecule, and the disconnected subgraph I have indicated above. As noted, it is conventional to regard the latter as a mere syntactic shorthand for the former, and accordingly, that any attempt to attach independent significance to the latter, to regard it as a separate *thing* apart from the whole big molecule which carries it, is a terrible mistake. If so, then that is precisely what the sequence hypothesis leads us to.

11F. Protein Folding and Morphogenesis

So far, we have couched our discussion entirely in terms of sequence; of primary structure, especially of polypeptides. Indeed, according to the sequence hypothesis, that is all DNA can encode; what it is information *about.* So let us take this seriously, and see where it leads us.

As we know, active protein molecules must have more than mere schematic structural formulas, more than primary structure. They must have shapes, forms, conformations of precise geometrical nature. For instance, here is a textbook representation of the enzyme ribonuclease (figure 11F.1).

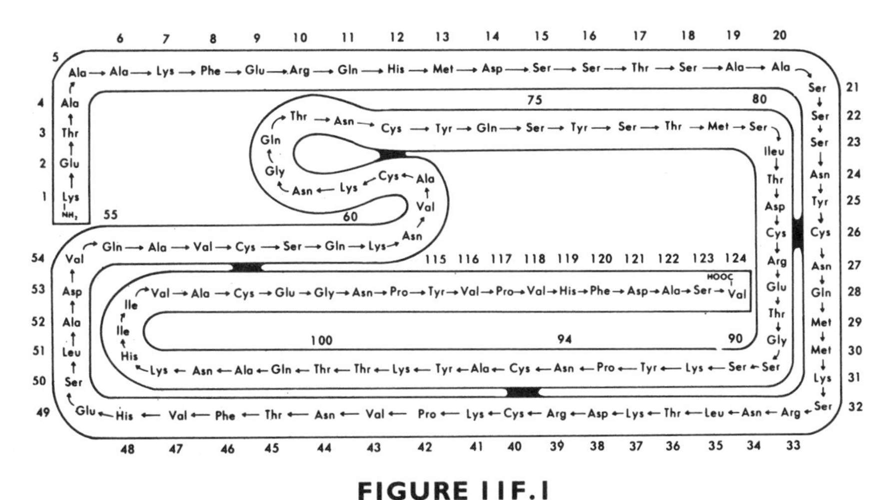

FIGURE 11F.1

From Smyth, Stein, and Moore, *J. Biol. Chem* (1963), 228:227.

This is very dramatic already, but it is still only a sketch; a two-dimensional projection. A three-dimensional representation is still more dramatic, as in figure 11F.2, a three-dimensional impression of the protein myoglobin.

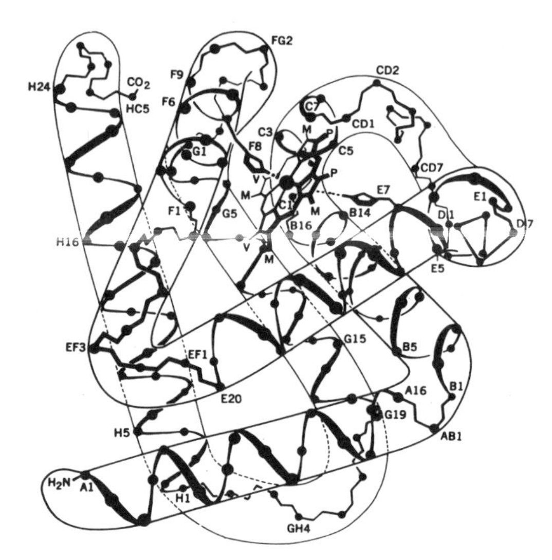

FIGURE 11F.2

From Neurath and Baley, *The Proteins,* 1962.

These drawings display the *tertiary* structure of the active molecules; the dramatic folds and bends which confer upon the molecule its characteristic shape, and hence its activities. (There is also a *secondary* structure, a regular geometry assumed by the molecule between these bends, and a *quarternary* structure, which pertains to multi-chain assemblies.) These shapes are not "coded" in DNA; as we have seen, the sequence hypothesis allows DNA to code only for primary structure. So where do these shapes come from? How does a primary structure clothe itself in a phenotype?

Indeed, there are two separate problems here: how do we get from primary structure to tertiary structure? and how do we get from tertiary structure to the actual activities, the functional roles of the molecules? How are these activities embodied in the shape?

The first of these is the *protein folding* problem. It is the problem of determining the relation of sequence (primary structure) to the active, folded *form* (i.e., tertiary structure), and even more, to characterize the morphogenetic process (folding) which generates that form. The second problem is even harder—to determine *activity* from form. We must find the "active sites," which let the molecule look into its environment, and the antigenic determinants, which allow it to be seen.

The basic fact is that folding is a spontaneous process. We can (gently) unfold an active protein, *denature* it, say by warming it. When restored to more physiological conditions, it will *refold* by itself, autonomously, very quickly and very accurately; all of its activities will be recovered. Thus it appears that tertiary structure somehow *inheres* in primary structure, that if you specify the sequence, you automatically get the folding.

These observations suggest that folding is *effective*. Church's Thesis (see 8.5 above) says that effective means *computable*. If so, we can solve the folding problem by producing an *algorithm* which generates tertiary structures from sequences. More precisely, we want an algorithm which will produce spatial coordinates of the individual residues of a folded molecule from the mere order of those residues along the molecule.

The strategy for arriving at such an algorithm seemed immediately clear from the strictures of physical chemistry, embodied in the concepts of *free energy* and *potential*. In a conservative system (the quintessential mechanism) we can associate with any configuration, in a particular set of environmental circumstances, a free energy; roughly, energy available for work, but which is not doing any. Associated with this is a *potential*, the negative of "total energy" minus "free energy." The basic law here is that configuration changes so as to lower free energy, and hence potential; in such circumstances, the dynamical laws governing temporal change of configuration are expressed by mandating that temporal rates of change of configuration be identified with appropriate partial derivatives (gradients) of this fixed potential. We thus move automatically to a configuration of *minimal* potential, *minimal* free energy; this is an *equilibrium* configuration, in which the temporal derivatives of configuration all vanish. And it is necessarily a *stable* equilibrium.

Molecules themselves are regarded as just such stable, minimal free-energy states, sitting at the bottoms of potential wells. In the context of protein folding, then, the active, folded form of a polypeptide chain must also be a minimal free-energy structure, driven to it from any initial (random) geometry. So all we need to do to get our algorithm is to write down the appropriate potential function, and then minimize it. More specifically, we must express this potential entirely in terms of primary structure under given environmental conditions; setting gradient of potential equal to rate of change of configuration gives the dynamics of folding, and minimizing the potential gives the corresponding folded forms directly.

The folding problem was early recognized as the outstanding problem of molecular biology, and the above strategy adopted from the very beginning (see, for instance, Harold Scheraga's *Protein Structure,* 1960). Though it was central to the entire picture we have sketched above, molecular biologists have tended to distance themselves from it, and hence trivialize it as something of no fundamental significance. Thus, Sydney Brenner says: "all you had to do was to specify the amino acid sequence and the folding would look after itself, and the energy would look after itself, and everything would be all right. Nothing to worry about. Now this was, I think, the blinding insight into the whole solution of everything in biology" (L. Wolpert and A. Richards, eds., *A Passion for Science,* 1988).

Actual experience, however, accumulated over the past three decades and more, does not bear out this rosy picture. From the outset, it was clear that such a "brute force" approach to folding would pose formidable technical problems at the very least. In physico-chemical terms, even a small polypeptide (ca. 100 amino acids) involves hundreds or thousands of atoms; writing down a potential function for these, even if nothing else is involved, is a Herculean task, and minimizing it is beyond any technical possibility. Thus, we must have recourse to metric approximations to the "real" potential, and believe that something folding in such an "approximate" potential field will fold approximately correctly.

This last, of course, is a stability assumption about folding dynamics—that it is insensitive to small perturbations, and hence may be replaced by an approximation to it. By adopting it, we can (and have) replaced the original strategy by a search for some technically feasible approximation; i.e., an approximation which remains a model.

Put baldly, thirty years of costly experience with this strategy has produced no evidence of this kind of stability. Despite a great deal of work by a lot of good people, working with ever more powerful hardware and ever more sophisticated software, the problem is still pretty much where it was in 1960. In fact, we are presently in a worse position than before, because we now do know, if not how proteins fold, that the obvious approaches do not work.

As noted earlier, the folding of real proteins is very fast and very accurate; it is quite effective. This must accordingly mean, according to the picture sketched above, that the "real" potential function (assuming there is one) must have a single deep minimum, a single well corresponding to the active conformation. On the other hand, the *approximations* with which people are working, that are presumably getting better in some sense as approximations, seem increasingly plagued by a plethora of relatively shallow local potential minima, in which configurations get trapped into inactive forms.

This is worse than being unsuccessful; it is simply contrary to experience. So we see explicitly here what we saw at the outset; when contemporary physics claims to speak about matters biological, it either has nothing to say, or it gives the wrong answers. And this is in an area where contemporary physics claims absolute authority; we are talking about *molecules,* not even about life at all yet.

The situation is rendered even more interesting by the following considerations. I have myself dabbled in protein folding, but from a morphogenetic rather than from a physico-chemical perspective. The origin of my approach was an apparently entirely unrelated group of morphogenetic phenomena called *cell sorting* (see G. D. Mostow, ed., *Mathematical Models of Cell Rearrangement,* 1975). Roughly speaking, we are concerned with the morphogenetic capabilities of populations of motile units, possessing differential affinities for each other. My contribution was to provide the basic models that answer such questions, at least phenomenologically. During a visit to the Soviet Union in 1972, I met the mathematician I. I. Pyatetskii-Shapiro, who had no particular concern with biology, but who found sorting congenial mathematically. Particularly, he had become interested in a problem he called *circularization*—the invention of sorting rules which would make one of the sorting populations assume the configuration of a regular polygon, however they were distributed initially. I saw in the proverbial flash how one could bring the morphogenetic ideas of cell sorting to bear on protein folding, by the simple expedient of tying one of the sorting populations together with inelastic string.

Experience with this approach has been most interesting. As with the physico-chemical approach, it proceeds by minimizing something; an *objective function.* But it is not the free energy of physical chemistry. Instead of the thousands of variables and parameters inherent in the latter approach, it contains very few; less than ten. But the interesting point is that among these ten or so control parameters, which manipulate the hundreds of spatial degrees of freedom of the folding polypeptide chain, some must incorporate "global" information, such as distance from a centroid. In turn, this information comes from a "generic" *folded* protein. That is, we must use the properties of such a "generic" folded protein as a model of a protein *to be folded.* This in turn may be regarded as creating an impredicativity, the hallmark of complex systems, and precisely the sort of thing which syntax alone cannot handle.

On the other hand, approaching folding from this direction reveals it to be a *synergetic* process; one in which very few controls can manipulate a much larger number of configurational degrees of freedom. Such synergies are everywhere in biology, as they also are in any inordinately constrained mechanism. My suggestion is, of course, that in biology they are indicators of complexity rather than of mechanisms under constraints.

Experience with folding models of this type has been very good by conventional standards, limited as it has been. By the criterion of comparison with crystallographic structures, even with crude models, we can obtain active conformations very quickly and very accurately. But insofar

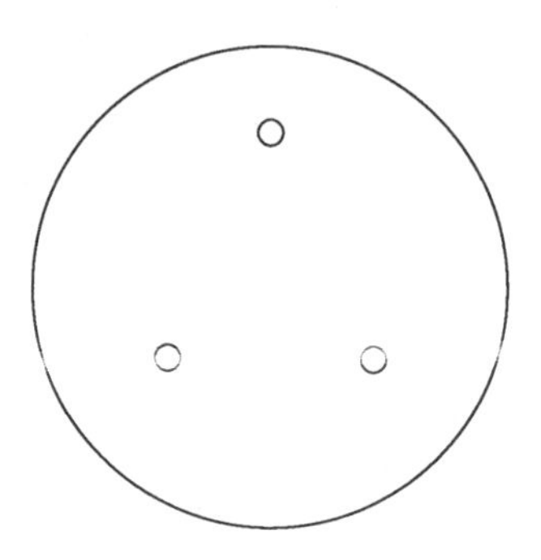

FIGURE 11F.3

as it embodies an aspect of complexity, in the very construction of the objective function, it is not even algorithmic in character at all; effective but not computable.

Let us now turn to the other problem we mentioned at the outset; namely, that of passing from active conformation to active sites, and hence to functional activity. Roughly speaking, folding serves to bring constituent residues that are remote in primary structure into close spatial proximity. Thus, in standard chemical terms, atoms and reactive residues are brought into, and held in, close spatial proximity, even though they seem far apart in terms of primary structure.

Heuristically, this is exactly what an "active site" is presumed to be. What I am going to argue now is that, although these "active sites" embody in themselves many of the properties of traditional chemical molecules, they are not molecules. Not being held together by internal chemical bonds of their own, they cannot be isolated as independent "substances"; as such, they are *not fractionable* in these terms from the bigger molecule which manifests them. They have sources from which they emerge, and sinks down which they disappear, but they are neither the products of conventional chemical reactions, nor are they used up thereby. Nevertheless, they actively participate in conventional molecular reactions, though which they can be characterized in *functional* terms. The reactants that interact with them can see them; indeed, that is *all* these reactants can see. But we have rendered them invisible to ourselves by our very way of intrinsically characterizing chemical structure. As such, they cannot be directly *coded for* via any purely syntactic scheme.

Indeed, this second problem, of going from primary structure to active site, manifests in a molecular microcosm the genotype-phenotype dualism we have already described above.

We have already pointed out that sequence or primary structure itself is

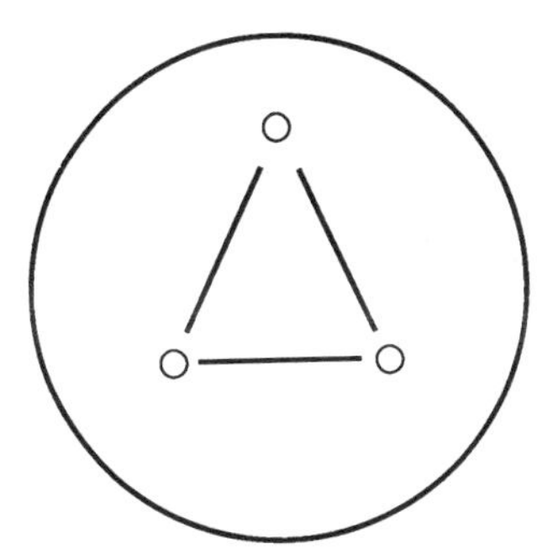

FIGURE 11F.4

embodied in disconnected subgraphs of the structural, graphical formulae of traditional chemistry. We are simply going to pursue these ideas a step further, in the context of active sites, to see how something can be *in* a molecule, and itself *act* like a molecule, but not *be* a molecule.

The basic idea here is very simple. Suppose we have a configuration of three "things," represented by the abstract vertices (figure 11F.3). If we want to maintain this configuration, to hold it together, we may do it in at least two different ways. The obvious thing is to tie these "things" together directly, by means of internal bonds; i.e., to connect them with *edges,* and turn the whole thing into a rigid structural *graph* (figure 11F.4)

A different, but quite equivalent possibility is not to use such internal bonds at all, but rather to hang our constituent things onto some larger, external structure which acts as scaffolding (figure 11F.5). In this last situation, the scaffolding itself may be a graph; if so, then the configuration it scaffolds is precisely a disconnected subgraph. Note that the scaffolding itself may be *quite arbitrary;* all it needs to do is to hold our configuration together.

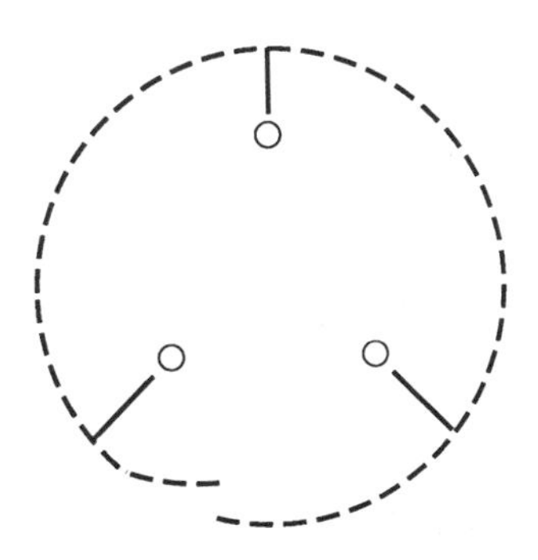

FIGURE 11F.5

As we have seen, sequence (e.g., of bases in DNA) pertains to scaffolding of this kind. Such sequences are held together, not by any direct intersymbol bonds, but by being suspended in a larger structure. Conversely, any larger structure that maintains their configuration would create the sequence; its exact nature, its chemistry, if you will, is otherwise irrelevant.

If we perchance interpret the elements of such configurations to be *atoms,* or chemical groups, or even bits and pieces of chemical groups, then such scaffolded configurations may themselves act like conventional chemical species. If so, they are in fact much more general than conventional molecules; there may be no way of holding them together through internal chemical bonds at all. They can only "exist" when scaffolded together in the way we have indicated. If the scaffolding as a whole is perturbed, or disrupted, they disappear, they cease to exist, they *denature.* But they do not "decompose," in any conventional sense, and they reappear when the scaffolding is restored.

What I am suggesting is that the *active sites* of folded polypeptides, or of multichain structures, are of this character. Hence that folding generates a scaffolding which, in this sense, brings entirely new chemical entities into existence, entities composed of parts drawn from residues remote from each other in terms of primary structure. Hence these scaffolded entities do not have a symbolic representation in terms of that structure at all; they are precisely the intersymbol hybrids mentioned earlier, which are forbidden in any syntactic scheme based on those symbols.

This kind of *site,* as a thing in itself, is an example of something nonfractionable from the scaffolding that carries it. Although it would take us far beyond the scope of the present discussion to demonstrate it, a corollary of this nonfractionability is that we cannot get at a functional description of the site from a purely structural characterization of the scaffolding. Thus, if the scaffolding is a conventional molecule, like a polypeptide chain, a description of the molecule does not yield a description of such a site; at least not in any conventional syntactic sense. Nor conversely, of course.

Many years ago, in a source I have long since forgotten, I read a comment of the preeminent organic chemist Richard Willstätter. He would not accept the identification of enzyme with protein, simply on the basis of the fact that enzymatic activity moved with protein in analytical procedures. He asserted rather that the protein was coming along with the enzymatic activity as an "unavoidable contaminant." The above considerations, especially regarding nonfractionability of site from scaffolding, should be considered as substantiating his claim.

What we have argued above is (1) there is no algorithm that will take us from primary structure to tertiary structure directly, and (2) there is further no algorithm that will take us from tertiary structure to functional activity, or "active sites." What is involved here is *complexity*, even here, at the molecular level, where there is as yet no life. Or, to state it another way: if phenotype is chemistry, as mandated by the sequence hypothesis, that "chemistry" is not the familiar contemporary chemistry we find in the textbooks. The species involved are not *synthesized* from elements in any ordinary sense; rather, they *emerge* through a process of morphogenesis. Thus, when we ask "why?" about them, when we treat them as effects and inquire into their causal correlates, we cannot make do with syntactic answers framed in terms of sequence. As always, there is not enough entailment in syntax alone to permit it.

11G. A Word on Entailment in Evolution

As we have just seen, the presumed machinelike, syntactic relation between genotype and phenotype in the biology of soma cannot be maintained. If the question "What is life?" pertains, as we have claimed, to soma and organism, then the proposed answer "Life is machine" is wrong.

In the present section, I shall briefly argue that if soma is not machine, then evolutionary processes are not devoid of entailment. That is: evolution is not to be regarded as a matter of unentailed "frozen accident," a matter of pure history of historical chronicle, and immune to Natural Law. Such a view can only be maintained if somatic biology is entirely a matter of software and hardware, with hardware unable to entail anything about its software (i.e., about its program and its initial data), and the program unable to change the hardware.

Perhaps the earliest biologically respectable (i.e., nontelic) assertion of entailment in evolutionary processes is found in Ernst Haeckel's idea of *recapitulation*, dating from 1866. This idea is usually expressed in the elliptical form "ontogeny recapitulates phylogeny," and often referred to as the "biogenetic law," or the "law of heteroauxesis." What it really asserts is that ontogeny, the morphogenetic processes which culminate in a particular organism, and phylogeny, the presumed evolutionary processes which culminate in that same organism, are in some sense *similar*. That is, they are intertransformable; alternate realizations of a common model. Haeckel's idea, in those very early evolutionary days (Darwin's *Origin of Species* had only been published in 1859) was to provide another line of evidence for evolution itself, and to suggest further that it might be studied through its embryological models.

Tacit in Haeckelian recapitulation is the idea that, if ontogenetic processes are subject to entailment, so are phylogenetic ones. It was not asserted that either phylogenetic processes entailed ontogenetic ones, or vice versa. But that is how people tried to interpret recapitulation, and the whole subject dissolved in a welter of controversy after a while.

A different version of exactly the same ideas is manifested in D'Arcy Thompson's venerable "Theory of Transformations," put forward in the last chapter of his epic *On Growth and Form,* first published in 1917. I paraphrase as follows: *closely related species are similar.* Here, "closely related" is a genotypic concept; "similar" pertains to phenotype. Hence we have here a highly nontrivial assertion, and one in genotype and phenotype, somas and germlines, enter simultaneously. I have discussed this matter extensively elsewhere (see *AS*) and will not repeat it here, save to mention that it is a manifestation of *structural stability.* It asserts that a sufficiently "small" but otherwise arbitrary perturbation of genome can be offset by a transformation (similarity) of phenotype alone.

The importance of this kind of proposition has perhaps never been sufficiently appreciated. For it underlies the universal supposition that we can learn about one species of organism (e.g., man) by acquiring data about another (e.g., a rat); that we can use one species (or even a specimen of that species) as a surrogate for another in phenotypic terms. And although people collect such surrogate data with great care, the conditions under which it is permissible, or even possible, to extrapolate from that data (i.e., to *transform* it into corresponding data pertaining to another species) receives no such attention. On the other hand, D'Arcy Thompson's principle cannot be universally true; if it were, there could be no such thing as *macroevolution.* That is: from this point of view, macroevolution requires *bifurcating genomes,* for which arbitrarily small changes can result in dissimilar (i.e., non-intertransformable) phenotypes.

To illustrate these ideas, let us examine a few of D'Arcy Thompson's own vivid and familiar diagrams, for instance, figures 11G.1 and 11G.2. In these figures, phenotype is explicitly represented; genotypes are manifested implicitly in the transformations that take us from one phenotype to another, transformations that depend uniquely on those genotypes.

As I noted earlier, such a transformation can be regarded as a vehicle for moving phenotypic "data" from one species to another, different one. But (and this is the crux for our purposes) we can equally well regard the transformation as determining the *genotype* of one of the species, given the two phenotypes and the genotype of the other. That is, we can *entail* one of the genotypes from the other, to the precise extent that the associated phenotypes can be brought into coincidence by a similarity transformation.

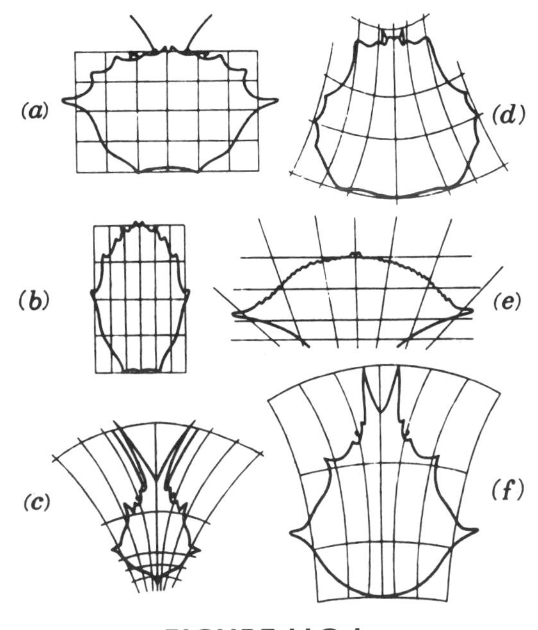

FIGURE 11G.1

Carapaces of various crabs: (a) *Geryon,* (b) *Corystes,* (c) *Scyramathia,* (d) *Paralomis,* (e) *Lupa,* (f) *Charinus.*

That similarity, it will be noticed, is a modeling relation that brings entirely somatic entailments into coincidence. So we can extend from these somatic entailments to corresponding entailments between the associated genomes.

As we see, this is precisely the sort of thing which Haeckel was claiming; i.e., that somatic entailments themselves entail evolutionary ones. But that is precisely what is forbidden in contemporary views about evolution (here, about macroevolution).

But what about bifurcating genomes, the province of macroevolution, where there are no similarity transformations between phenotypes? These can roughly be regarded as degenerate situations, where otherwise distinct chains of entailment coalesce. As we have seen before, in such degenerate situations, the behavior arising from a perturbation depends on *how* it is perturbed. It is only here that we have even the appearance of "historical accident," because we place the source of the perturbation *outside* the entailments in the system. That is, we put missing entailments into the environment, where everything is unentailed.

Oddly enough, if we suppose that such bifurcating genomes are in some sense *rare,* or sparse, in the set of all genomes, we can invoke the argument of Elsasser that we mentioned at the outset. To illustrate, let us

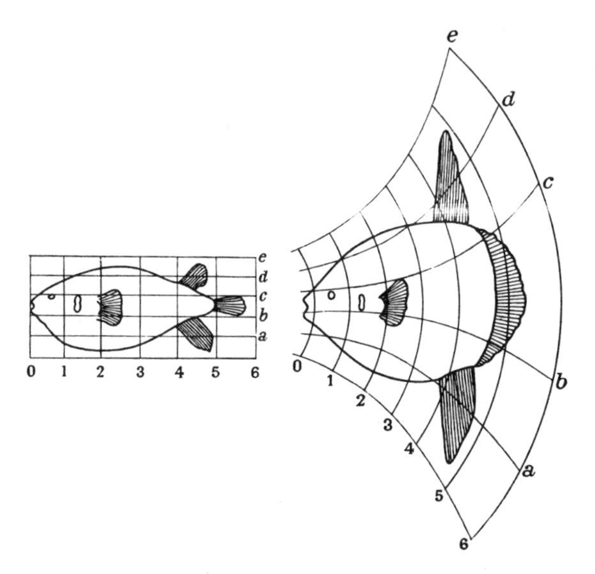

FIGURE 11G. 2

consider a simple physical example of such bifurcation, the state transition from solid to fluid. This is an example of what Thom (*Structural Stability and Morphogenesis,* 1975) called a cusp catastrophe. Near the cusp point, which is as rare as anything (there is only one), there are two minimal fold lines (also sparse) emanating from it. There is nothing in the system to decide which of these two minima should be chosen. Rather, there is an *extra condition,* which Thom called "Maxwell's convention," to lift the degeneracy. This extra condition, which otherwise seems to devolve on extraneous history, is a precise example of what Elsasser called a "biotonic law," something specifically pertaining to a sparse set; it is "extra information" that serves to internalize a mode of entailment that would otherwise have to be assigned to an acausal environment.

The entailments involved in Maxwell's convention become meaningless off this sparse set of bifurcation points. But on that set, they provide precisely the entailments that are lost through the degeneracy. The degeneracy, in turn, is what makes the special, nongeneric bifurcation points special in the first place; to characterize them as special requires extra conditions (i.e., the Maxwell convention). In this context, Monod's objections to Elsasser's arguments translate into calling Maxwell's convention vitalistic, because it "adds itself" to what happens on a special set. In fact, all it does is allow us to make better models, by pulling into the system relevant entailments we would otherwise have to leave "outside."

Just to close the historical circle, it may be noted that one of Thom's major sources of inspiration was D'Arcy Thompson's book; Thom describes his own book as "attempt(ing) a degree of mathematical corroboration" of those ideas.

11H. Relational Biology and Its Realizations

In this final section, we come back to our original question, which has animated the entire enterprise: What is life? In the last analysis, biological ideas must be assessed in terms of how well they serve at answering this question. No other factors are relevant.

As we have seen, contemporary biology gives two kinds of answers to this question. In somatic terms, the answer is: life is machine, a purely syntactic device, a gadget, to which a reductionistic strategy may be universally applied. In evolutionary terms, on the other hand, life is what evolves; the evolutionary process itself, which takes us from gadget to gadget, is devoid of entailment, the province of history and not of science at all.

To me, neither of these answers, either separately or together, serves to answer the question. If somatically an organism is a machine to be understood in purely syntactic, reductionistic terms, then life is only a matter of putting its fractions back together. But as we all know, it is literally not that simple, not even when one admits randomness, the antithesis of mechanism, as the ultimate aspect of somatic organization, or self-organization. And evolution, entailed or not, has from the beginning concerned itself only with origin of *species* of life; it does not bear on life itself.

As we have seen, there are many good reasons for *wanting* to be a reductionist, but unfortunately these have nothing to do with answering the question. One reason is that reduction, syntax, has since the time of Newton become identified with science itself, and that any deviation from its prescribed algorithmic progression from earlier to later, any shred of function or finality, any manifestation of semantics, is mysticism. Syntax itself has become confounded with what is *objective,* and this objectivity is what makes science a democratic activity, something which anybody can do, by simply following an algorithm. And, of course, reductionism carries with it the lure of *unification;* of having to know ultimately only one thing, one principle, from which everything else syntactically follows. All these things are very attractive, but as noted, they have nothing much to do with answering a question. They provide only a possible strategy for getting answers.

And yet, as Einstein kept insisting, science involves a free creative act of their intellect; ultimately, it involves wisdom. It involves the ability to select what is important about a problem from what is irrelevant or incidental, and to follow that. There is no algorithm for this, just as there is no algorithm for making a model. It would perhaps be nice if there were such algorithms, which would make wisdom irrelevant, and indeed put the wise and unwise on an equal footing. But there are not and, as Herodotus remarked long ago: "Where wisdom is required, force is of little avail." So what is the alternative?

One clue to the alternative is embodied in the observation of Rashevsky (see section 5B above) that no (finite) concatenation of syntactic models of an organism yields something which must be an organism. Or, to put it otherwise, every concatenation of such models can be realized by something which is not alive. We do not seem to be able to stop at any finite point in this modeling process and say that we are done. From this, it is only a step to realizing that something is special about material systems whose properties can thus be syntactically exhausted at some finite point, a point at which we have reached a "largest model." Rashevsky's observation suggests that organisms are not in this class of systems; as we have seen, this is one way into the world of *complex* systems, systems that have no such largest syntactic model.

As we have seen, a corollary is that complex systems cannot be exhausted by reductionistic fractionation either. Just as we cannot concatenate syntactic models to obtain an organism, we cannot, for the same reason, concatenate reductionistic fractions to get an organism. I have discussed these matters at great length above, and need not repeat the discussion here.

But complexity, though I suggest it is the habitat of life, is not itself life. Something else is needed to characterize what is alive from what is complex. Rashevsky provided this too, in his idea that biology was *relational,* and that relational meant (as we stated it) throwing away the physics and keeping the organization. A rough analog would be: throwing away the polypeptide and keeping the active sites. Organization in its turn inherently involves *functions* and their interrelations; the abandonment of fractionability, however, means that there is no kind of 1 to 1 relationship between such relational, functional organizations and the structures which realize them. These are the basic differences between organisms and mechanisms or machines.

I was perhaps fortunate in having early experience with systems of entailment that are complex in this sense. As I noted at length above, mathematics is replete with them. So to me, it involved no great shock to depart from the fold of mechanism, if that was what the problem required; doing so, to me, was not a departure from science, or from entailment itself. In a historical context, it is radical, but that is irrelevant. The only pertinent circumstance is whether we can better address the question this way. I believe we can, and I have set out my arguments for this belief at length above. I of course find them persuasive, but that is for the reader to assess.

Index